Handbook of
Basic Electrical Engineering Formulae

Harish C Rai

PhD (Electrical Engg, IIT, Delhi), FIE (India), FIETE, MISTE, MAeSI

Pro Vice Chancellor
Galgotias University, Greater Noida, UP

Former
Professor, Department of Electrical and Electronics Engineering
Chhotu Ram State College of Engineering
(Presently Deenbandhu Chhotu Ram University of Science and Technology)
Murthal, Haryana 131039

Controller of Examinations
Director, Academic Affairs
Director, Research Project Monitoring Cell
Director, Organization and Development
GGS Indraprastha University, Delhi

Advisor I, All India Council of Technical Education
(Ministry of Human Resource and Development, Delhi)

Shivanshu Rai

Project Manager
Tata Consultancy Services (USA)

CBS Publishers & Distributors Pvt Ltd

New Delhi • Bengaluru • Chennai • Kochi • Kolkata • Mumbai
Hyderabad • Jharkhand • Nagpur • Patna • Pune • Uttarakhand

Handbook of
Basic Electrical Engineering Formulae

ISBN: 978-93-87085-00-8

Copyright © Authors and Publisher

First Edition: 2018

Published by Satish Kumar Jain and produced by Varun Jain for

CBS Publishers & Distributors Pvt Ltd

4819/XI Prahlad Street, 24 Ansari Road, Daryaganj, New Delhi 110 002, India.
Ph: 23289259, 23266861, 23266867   Website: www.cbspd.com
Fax: 011-23243014   e-mail: delhi@cbspd.com; cbspubs@airtelmail.in.
Corporate Office: 204 FIE, Industrial Area, Patparganj, Delhi 110 092
Ph: 4934 4934   Fax: 4934 4935   e-mail: publishing@cbspd.com;
publicity@cbspd.com

Branches

- **Bengaluru:** Seema House 2975, 17th Cross, K.R. Road,
  Banasankari 2nd Stage, Bengaluru 560 070, Karnataka
  Ph: +91-80-26771678/79   Fax: +91-80-26771680   e-mail: bangalore@cbspd.com
- **Chennai:** 7, Subbaraya Street, Shenoy Nagar, Chennai 600 030, Tamil Nadu
  Ph: +91-44-26680620/26681266   Fax: +91-44-42032115   e-mail: chennai@cbspd.com
- **Kochi:** Ashana House, No. 39/1904, AM Thomas Road, Valanjambalam, Ernakulam 682 016,
  Kochi, Kerala
  Ph: +91-484-4059061-65   Fax: +91-484-4059065   e-mail: kochi@cbspd.com
- **Kolkata:** 6/B, Ground Floor, Rameswar Shaw Road, Kolkata-700 014, West Bengal
  Ph: +91-33-22891126, 22891127, 22891128   e-mail: kolkata@cbspd.com
- **Mumbai:** 83-C, Dr E Moses Road, Worli, Mumbai-400018, Maharashtra
  Ph: +91-22-24902340/41   Fax: +91-22-24902342   e-mail: mumbai@cbspd.com

Representatives

• **Hyderabad**	0-9885175004	• **Nagpur**	0-9021734563
• **Patna**	0-9334159340	• **Pune**	0-9623451994
• **Uttarakahnd**		• **Jharkhand**	

Printed at: India Binding House, Noida, UP, India

to
My respected parents
Late Sh Balraj
Late Smt Ram Devi
My lively children, Shivanshu and Himanshu
My loving wife, Sangeeta;
and Shipra and grandson Vivaan, whose
affection is always appreciated

Harish C Rai

Preface

The purpose of this book is to serve as a reference for practising electrical engineers, students, teachers and supervisors. The material has been carefully compiled so that it can serve the needs of students and professionals who wish to have a ready-reference to formulae, equations, methods, concepts and their mathematical formulations.

Chapters 1–11 cover mathematical formulae and concepts used by most electrical engineers with major emphasis on topics frequently occurring in the solution of physical problems. Chapters 12–24 encompass a wide range of subjects that include the basics of electrical engineering; electromagnetics; circuit theory and network; electrical machines; measuring instruments; control engineering; electrical power generation, transmission and distribution; protection; power electronics; logic circuits; communication and mapping. Electrical engineers, with the help of this book will be able to tackle their day to day field problems with confidence. The book, which gives an integrated treatment of the entire subject, is intended to fill this need.

Appendices at the end are a culmination of tables of numericals of the most important functions. The book will be useful to both college and university students as well as engineers working in the industry.

It is hoped that this book will be of immense use to teachers and students of technical institutes. Suggestions from students, teachers and working engineers for improvement in future editions of this book are welcome.

Harish C Rai
Shivanshu Rai

Acknowledgements

I thank all my undergraduate students who suggested that I should write this book and indeed, all those who have encouraged me in this venture. I derive immense pleasure in expressing my sincere thanks to Prof Yogesh Singh, Vice Chancellor, and Prof Annu Singh Lather, Pro Vice Chancellor, Delhi Technological University (DTU), for the invaluable encouragement throughout this work. I am indebted to their guidance and invaluable suggestions.

I express my gratitude to Prof SS Murthy, former Vice Chancellor, Central University of Karnataka; Prof ZH Zaidi, former Vice Chancellor, MJP Rohilkhand University, Bareilly; Prof BP Singh, Prof Bhim Singh, Department of Electrical Engineering, IIT, Delhi, for sparing their valuable time and providing useful guidance on various chapters.

I thank my colleagues, Prof Alok Mittal, Member Secretary, AICTE, New Delhi; Prof JRP Gupta (NSIT, Delhi); Prof DR Bhaskar (DTU); Prof SS Inamdar (Vishwaniketan, Mumbai); Prof VK Sharma (NIT, Uttarakhand); Prof Rominder Randhwa, Director, Guru Tegh Bahadur Institute of Technology; Prof SS Tyagi, Director, BSA Institute of Technology, Haryana, and Prof Lajpat Rai, IIT, Delhi with whom I have discussed power electronics while teaching courses on this subject.

I express my gratitude to my brother Dr Mahesh Popli (Income Tax Department), Rajasthan; Dr Vikas Gupta (DU); Mr Pankaj Munjal, Director, Training and Development, RVIT, Bijnore; Brig Pradeep Upmanu; Dr Nitin Malik, GGS Indraprastha University, and Mr Ankit Popli for their immense help and constructive criticism on the manuscript.

My special thanks are due to Sh RC Taneja and late Sh KR Munjal for their moral support which has enabled me to complete this work. I am grateful to Sh Satish Kumar Jain (Mataji), CMD, and Sh Varun Jain, Director, CBS Publishers & Distributors,

New Delhi, for their patience, goodwill and cooperation. I express my gratitude to Mr YN Arjuna (Senior Vice President Publishing, Editorial and Publicity); Mrs Ritu Chawla (AGM Production); Mr Sumit Behl; Ms Sanjubala Tripathy (Copy Editor) and Mr Parmod Kumar (DTP Operator) for bringing out the book in the present form.

Finally, I appreciate the patience and solid support of my family—my wife Sangeeta Rai; children Shivanshu, Shipra and Himanshu.

Harish C Rai

Contents

Chapter 1

Useful Tables

1. Weights and Measures

Linear measure

4 inches	= 1 hand	= 10.16 cm	6 feet	= 1 fathom	
9 inches	= 1 span	= 28.86 cm		= 1.828 metre	
12 inches	= 1 foot	= 30.48 cm	22 yards	= 1 chain	
3 feet	= 1 yard	= 0.914 m	1 metre	= 39.37 inches	
5 feet	= 1 pace	= 1.524 m	10 chains	= 1 furlong	
			8 furlongs	= 1 mile	

Cubic or solid measure

Cubic foot = 1728 cubic inches × 16.387 = 28317 cubic centimetres

Cubic yard = 27 cubic feet = 21.033 bushels = 0.7645 cubic metre

Shipping ton = 40 cubic feet of merchandise = 1.13 cubic metre

Shipping ton = 42 cubic feet of timber = 1.18 cubic metre

One ton or load = 50 cubic feet of hewn timber = 1.42 cubic metre

One ton of displacement of a ship = 35 cubic feet = 1.02 cubic metre

Square or land measure

144	square inches	= 1 square foot
9	square feet	= 1 square yard
1210	square yards	= 1 rood
4	roodes	= 1 acre (0.407 hectare)

$$640 \ \text{acres} \qquad = 1 \ \text{square mile}$$

$$1 \ \text{square link} \quad = 62\tfrac{3}{4} \ \text{square inches (approx)}$$

$$1 \ \text{square chain} = 10{,}000 \ \text{square links} = 484 \ \text{square yards}$$

$$33 \ \text{square yards} = 1 \ \text{rod of building} = 27.6 \ \text{square metres}$$

$$100 \ \text{square feet} \quad = \text{Square of flooring or roofing} = 9.3 \ \text{sq. metres}$$

$$272\tfrac{1}{2} \ \text{square feet} \quad = \text{Rod of brick layer's work} = 25.4 \ \text{sq. metres}$$

Avoirdupois weight

16 drums	=	1 oz (437.5 grains)	28 lb	= 1 qr
16 ounces	=	1 pound (lb)	112 lb	= 1 cwt
14 pounds	=	1 stone	20 cwt	= 1 ton

Fluid memoranda

1 cubic foot of water $= 6\tfrac{1}{4}$ gals (approx) $= 6\tfrac{1}{2}$ lb $= 7.48$ US gals

1 US gal = 231 cubic inches = 0.1337 cubic ft
1 lb water at $62\,°F = 0.016$ cub ft
1 BI gal = 277.418 cub inch; 1 cwt of water = 1.8 cubic ft = 11.2 gals
1 British gal = 1.2009 US gals; 1 ton of water = 35.9 cubic ft = 224 gals
1 inch of rainfall = 22.622 gals per acre = 100 ton (approx)

Substance	lb/gal	Substance	lb/gal
Acetic acid	10.49	Petrol	7.50
Alcohol	8.00	Sperm oil	8.80
Hydraulic acid	12.00	Sulphuric acid 98%	18.35
Mercury	135.90	Turpentine	8.70
Milk	10.30	Water (distilled)	10.00

2. Greek Symbols

Symbol			Notation	Symbol			Notation
alpha	A	α	angle, coefficients	mu	M	μ	bending moment,
beta	B	β	angle, coefficients				coefficient of friction,
gamma	Γ	γ	specific gravity				permeability

(Contd...)

delta	Δ δ	density, increment, finite diff. operator	nu	N ν	Kinematic viscosity, frequency, reluctivity
epsilon	E ε	hyper log., linear strain, permittivity	Xi omincron	Ξ ξ O o	output coefficient
zeta	Z ζ	coordinate, coefficient	pi	Π π	circum., ÷ diameter
eta	H η	magnetic field strength, efficiency	rho sigma	P ρ Σ σ	specific resistance summation
theta	Θ θ	angular disp. time	tau	T τ	time constant
iota	I *i*	inertia	upsilon	Y υ	
kappa	K κ	bulk modulus, mag- netic susceptibility	phi chi	Φ φ X ξ	flux, phase
lambda	Λ λ	permeance, conduct, wavelength	psi omega	Ψ ψ Ω ω	angular velocity angular velocity

3. Physical Constants

Electron charge (e)	$= 1.602 \times 10^{-19}$ coulomb
Electron rest mass (m_e)	$= 9.11 \times 10^{-31}$ kg $= 0.511$ MeV
Electron (charge/mass) (e/m_e)	$= 1.760 \times 10^{11}$ coulomb/kg
Avogadro constant (N_A)	$= 6.023 \times 10^{23}$ per mole
Boltzmann constant (k)	$= 1.38 \times 10^{-23}$ J/°K
Faraday constant (F)	$= 9.65 \times 10^{4}$ coulomb/mole
Gas constant (R)	$= 8.31 \times 10^{3}$ J·°K^{-1}·kmole^{-1}
Gas (ideal) normal volume (V_0)	$= 22.4$ m^3/kmole
Gravitational constant (G)	$= 6.67 \times 10^{-11}$ N·m^2/kg
Mass of proton (rest mass) (m_p)	$= 1.673 \times 10^{-27}$ kg $= 938.8$ MeV
Mass of hydrogen atom (rest mass) (m_H)	$= 1.673 \times 10^{-27}$ kg $= 938.8$ MeV
Mass of neutron (rest mass) (m_n)	$= 1.675 \times 10^{-27}$ kg $= 939.6$ MeV
Plank's constant (h)	$= 6.63 \times 10^{-34}$ J. sec
Speed of light (c)	$= 3 \times 10^{8}$ m/sec
Melting point of ice	$= 0\,°C$ or $273.15\,°K$

$$\pi = 3.14159, \quad e = 2.71828$$
$$\log_{10} e = 0.43329, \quad \log_e 10 = 2.30258$$

4.   Prime Numbers

2	3	5	7	11	13	17	19	23	29
31	37	41	43	47	53	59	61	67	71
73	79	83	89	97	101	103	107	109	113
127	131	137	139	149	151	157	163	167	173
179	181	191	193	197	199	211	223	227	229
233	239	241	251	257	263	269	271	277	281

Table 1.1   Physical quantities and their units

Physical Quantity	Name of Unit	Symbol for SI Unit
Length	metre	m
Mass	kilogram	kg
Time	second	s
Electric current	ampere	A
Thermodynamic temperature	kelvin	°K
Luminous intensity	candela	cd
Amount of substance	mole	mole
Plane angle	radian	rad
Solid angle	steradian	sr

Table 1.2   SI units

Physical Quantity	SI Unit	Unit Symbol	
Force	newton	N	$= \text{kg m/s}^2$
Work, energy, quantity of heat	joule	J	$= \text{N} \cdot \text{m}$
Power	watt	W	$= \text{J/s} = \text{N} \cdot \text{m/s}$
Electric charge	coulomb	C	$= \text{A} \cdot \text{s}$
Electric potential	volt	V	$= \text{W/A}$
Electric capacitance	farad	F	$= \text{A} \cdot \text{s/V}$
Electric resistance	ohm	Ω	$= \text{V/A}$
Magnetic flux	weber	Wb	$= \text{V} \cdot \text{s}$
Inductance	henry	H	$= \text{V} \cdot \text{s/A}$
Luminous flux	lumen	lm	$= \text{cd} \cdot \text{sr}$
Illumination	lux	*lx	$= \text{lm/m}^2$
Radiation: dose	rad	rad	
exposure	rongtan	R	
activity	curie	C_i	

* sr = steradian

Table 1.3   Physical quantities with unit symbol

Physical Quantity	SI Unit	Unit Symbol
Area	square metre	m^2
Volume	cubic metre	m^3
Frequency	cycle per second	s^{-1}
Density	kilogram per cubic metre	kg/m^3
Velocity	metre per second	m/s
Angular velocity	radian per second	rad/s
Acceleration	metre per second square	m/s^2
Angular acceleration	radian per second square	rad/s^2
Pressure	newton per square metre	N/m^2
	bar	$10^5 N/m^2$
Surface tension	newton per metre	N/m
Dynamic viscosity	newton second per metre square	$N \cdot s/m^2$
Kinetic viscosity (diffusion coefficient)	metre square per second	m^2/s
Thermal conductivity	watt per metre kelvin	W/mK
Electric field strength	volt per metre	V/m
Magnetic flux density	weber per square metre	Wb/m^2
Magnetic field strength	ampere per metre	A/m
Luminance	candela per square metre	cd/m^2

Table 1.4   General conversions

Multiply by	To Obtain	From	Multiply by
	To convert	To	
2.54	inches	centimetres	.3937
30.48	feet	centimetres	.0328
.914	yard	metres	1.094
1,609.3	miles	metres	.000621
1,853.27	nautical miles	metres	.000539
6.45	square inches	sq centimetre	.155
.093	square feet	sq metres	10.764
.836	square yard	sq metres	1.196
16.39	cubic inches	cubic centimetre	.061
28.3	cubic feet	litre	.0353
6.24	cubic feet	gallons	.1602
.765	cubic yard	cubic metre	1.308
.3732	pound (troy)	kilogramme	2.68

(Contd...)

Table 1.4 General conversions (*Contd.*)

	To obtain	From	Multiply by
Multiply by	To convert	To	
31.10	ounces (troy)	grammes	.03216
.4536	pound (avoir)	kilogrammes	2.2045
7,000	pounds (avoir)	grains (troy)	.000143
28.35	ounces (avoir)	grammes	.0352
.065	grain	grammes	.1538
50.8	cwt.	kilogrammes	.01968
1,016.0	tons	kilogrammes	.000984
4.546	gallons	litres	.22
10	gallons of water	pounds	.1
.454	pound of water	litres	2.202
70.3	lb per sq m	gm/sq cm	.0142
2.3	lb per sq in	head of water (ft)	.434
.7	lb per sq in	head of water (m)	1.4285
.068	lb per sq in	atmospheres	14.7
1.575	tons per sq in	kgm/sq mm	.635
4.883	lb per sq ft	kgm/sq metre	.205
.593	lb per cub yd	kgm/cub metre	1.686
16.02	lb.per cub ft	kgm/cub metre	.0624
.0998	lb. per gallon	kgm/litre	10.02
.138	foot lb	kgm·m	7.23
.33	foot tons	tonne metres	3
1.014	horse power	force de cheval	.9861
746	horse power (British)	watts	.00134
33,000	horse power	ft·lb/min	1/33000
76	horse power	kg·m/sec	.01316
44	watts	ft·lb/min	.0227
0.1	watts	kg·m/sec	10
0.252	B·Th·U	kg calories	3.97
14.7	atmospheres	lb/sq inch	.068
0.90	German candles	English candles	1.1111
9.55	carcels	candle	0.1047
.737	joules	ft·lb	1.357
88	mile/hour	ft/min	.01136
197	metre/sec	ft/min	.00508
1.8	C·H·U	B·Th·U	.5555
.0000208	centipoise	lb force sec sq ft	48000

Table 1.5 Conversion factors

S. No.	To obtain	From	Multiply by
1	Angstrom	m	10^{-10}
2	Atmospheres	kg/m^2	10332
3	Bars	kg/m^2	1.02×10^4
4	BTU	Joule	1054.8
5	BTU	kWh	2.928×10^{-4}
6	Circular miles	m^2	5.067×10^{-10}
7	Cubic feet	m^3	0.02831
8	Dyne	newton	10^{-5}
9	Erg	Joule	10^{-7}
10	Erg	kWh	0.2778×10^{-13}
11	Gauss	tesla	10^{-4}
12	Grams (force)	newton	9.807×10^{-3}
13	Horse Power (metric)	watts	735.5
14	Lines/sq. inch	tesla	1.55×10^{-5}
15	Maxwell	weber	10^{-8}
16	Mho	siemens	1
17	Micron	metre	10^{-6}
18	Miles	km	1.609
19	Miles	cm	2.54×10^{-3}
20	Poundals	newton	0.1383
21	Pounds	kilograms	0.454
22	Pounds (Force)	newton	4.448

Table 1.6 Standard wire gauge (SWG)

SWG	Diameter (mm)	Area (mm^2)	SWG	Diameter (mm)	Area (mm^2)
1	7.62	45.6	21	0.813	0.519
2	7.01	38.6	22	0.711	0.397
3	6.40	32.2	23	0.610	0.292
4	5.89	27.3	24	0.559	0.245
5.	5.38	22.8	25	0.508	0.203

(Contd...)

Table 1.6 Standard wire gauge (SWG) (*Contd...*)

SWG	Diameter (mm)	Area (mm²)	SWG	Diameter (mm)	Area (mm²)
6	4.88	18.7	26	0.457	0.164
7	4.47	15.7	27	0.417	0.136
8	4.06	13.0	28	0.376	0.111
9	3.66	10.5	29	0.345	0.0937
10	3.25	8.3	30	0.315	0.0779
11	2.95	6.82	31	0.295	0.0682
12	2.64	5.48	32	0.274	0.0591
13	2.34	4.29	33	0.254	0.0507
14	2.03	3.24	34	0.234	0.0429
15	1.83	2.63	35	0.213	0.0357
16	1.63	2.07	36	0.193	0.0293
17	1.42	1.59	37	0.173	0.0234
18	1.22	1.17	38	0.152	0.0182
19	1.02	0.811	39	0.132	0.0137
20	0.914	0.657	40	0.122	0.0117

Table 1.7   Specific resistance (ρ) of common materials at 20 °C

Material	ρ (ohm·m)	Material	ρ (ohm·m)
Aluminium	2.83×10^{-8}	Lead	22×10^{-8}
Brass (annealed)	7×10^{-8}	Manganin	45×10^{-8}
Constantan	49×10^{-8}	Nichrome	109.7×10^{-8}
		Nickel	7.81×10^{-8}
Copper (annealed)	1.724×10^{-8}	Platinum	10×10^{-8}
Copper (hard drawn)	1.78×10^{-8}	Silver	1.64×10^{-8}
Gold	2.44×10^{-8}	Steel (hard)	4.57×10^{-8}
Iron (commercial)	12×10^{-8}	Tungsten	5.52×10^{-8}
		Zinc	6.22×10^{-8}

Table 1.8 Temperature limits of solid dielectrics

Class	Maximum Temperature	Example
Y	90 °C	Cotton, silk, wood, paper, fibre without impregnation, etc.
A	105 °C	Black tape, empire cloth and sleeves, impregnated paper, varnished silk
E	120 °C	PVC, synthetic plastic, synthetic resin, cotton and paper, laminates with formaldehyde bonding mica paper
B	130 °C	Combination of mica, tape, glass, fibre, asbestos with suitable binding
F	155 °C	Mica sheets, micanite with cement glass fibre, asbestos fibre, impregnated and dried paper
H	180 °C	Silica based inorganic and organic compounds, glass fibre and asbestos material with silicon resins
C	above 180 °C	Mica, porcelain, glass, quartz, asbestos, magnesium oxide

Table 1.9 Dielectric strength

Material	Dielectric Strength (kV/mm)	Material	Dielectric Strength (kV/mm)
Air	3	Mica	50–200
Asbestos	10	Paper	4–10
Bakelite	10–28	Paper (kraaft)	20–30
Bitumin (vulkanised)	14	Paraffin wax	8
Cotton	3–4	Porcelain	7–20
Empire cloth	10–20	Rubber	10–25
Fibre	6	Shellac	5–20
Glass	10–40	Varnished cambric	32
		Vinyl (plastic)	16

Table 1.10 Relative permittivity

Material	Relative Permittivity ε_r	Material	Relative Permittivity ε_r
Vacuum	1	Paper	2–4
Air	1.006	Paraffin	2–2.5
Bakelite	4.5–5.5	Plastic	2–4.5
Ceramic (low loss)	6–20	Porcelain	6
Ceramic (high loss)	more than 1000	Rubber	2–3
Fibre	2–5	Shellac	2–5
Glass	5–100	Transformer oil	4
Mica	3–7	Wood	2.5–7

Table 1.11 Illumination parameters, relationships and units in lighting engineering

Physical Quantity	Symbol	Relationship (Simplified)	SI Unit	Explanation
Luminous flux	Φ		Lumen or lm	Photometrically evaluated radiant flux (light output)
Luminous intensity	I	$I = \dfrac{\Phi}{\Omega}$	Candela, cd	Quotient of luminous flux Φ, emitted from a light source in a specific direction and the solid angle Ω
Illuminance	E	$E = \dfrac{\Phi}{A}$ $E = \dfrac{I_e}{r^2}\cos^5 \varepsilon$	Lumen per square m	Quotient of luminous flux Φ, hitting surface A and, the illuminated surface A
Illumination	E	$E = \dfrac{I_e}{h^2}\cos^5 \varepsilon$	Lux	A surface in m^2 r distance in m h distance of light source to surface in m

(Contd...)

Table 1.11 Illumination parameters, relationships and units in lighting engineering (*Contd.*)

Physical Quantity	Symbol	Relationship (Simplified)	SI unit	Explanation
Luminance	L	$L = \dfrac{\Phi}{A\cos\varepsilon\,\Omega}$ or $= \dfrac{I_e}{A\cos\varepsilon}$	Candela per square metre (cd.m^{-2})	ε radiation angle of lamp or incidence angle on surface measured to vertical I_e luminous intensity at radiation angle Quotient of luminous flux Φ, passing through surface A in a specific direction ε and the product of the solid angle Ω and the projection of surface $A \times \cos\varepsilon$ on a plane vertical to the viewed direction

Table 1.12   Lighting calculation for indoor working spaces (according to efficiency method)

$n = \dfrac{1.25\,EA}{\Phi_L \cdot \eta_B}.$	n	Number of Lamps Required
	1.25	The factor by which the design value should be chosen to exceed the nominal illuminance E_n
	A	Working surface in m^2
	Φ_L	Luminous flux of lamp in 1m
	η_B	Utilization factor

Utilization factors

Type of Lighting	η_B^*	Type of Lighting	η_B^*
Direct	0.6 to 0.45	Semi-indirect	0.45 to 0.35
Semi-direct	0.55 to 0.45	Indirect	0.35 to 0.25
Direct-indirect	0.5 to 0.35	Indirect cornice lighting	0.2 to 0.15

*According to room shape, as well as reflection factors of walls, ceilings and floors.

Table 1.13 Commonly used light sources, general purpose lamps—220 to 230 V

Wattage	Luminous Flux at 225 V	Base	Wattage	Luminous Flux at 225 V	Base
W	lm		W	lm	
25	230	E 27	300	5000	E 40
40	430	E 27	500	8400	E 40
60	730	E 27	1000	18800	E 40
75	960	E 27			
100	1380	E 27			
150	2250	E 27			
200	3150	E 27			

Table 1.14 Fluorescent lamp—Osram lumilux (with fully electronic control gear)

Wattage without Ballast	Wattage with Ballast	Rated Current	Length of Lamp	Luminous Flux According to Type of Lamp and Light Colour	
W	W	A	mm	lm	
16 (L 18 W)	19	0.095	590	1350	
32 (L 36 W)	36	0.18	1200	3200	
34 (L 38 W)	38	0.19	1047	3150	
50 (L 38 W)	55	0.25	1500	5200	
Fluorescent Lamps (with ballast)				Standard	Lumilux
8	14	0.145	288	330	430
15	25(19.5)*	0.33	438	720	1000
18	30(23)*	0.37	590	1150	1450
20	32(26)*	0.37	590	1150	–
30	40	0.365	895	1800	2400
36	46	0.43	1200	3000	3450
38	50*	0.43	1047	–	3200

(*Contd...*)

*For each lamp with series connection of two lamps to ~220 V and one ballast.

Table 1.14 Fluorescent lamp—Osram lumilux (with fully electronic control gear) (*Contd.*)

40	50	0.43	1200	3000	–
58	71	0.67	1500	4800	5400
65	78	0.67	1500	4800	–

*With 40 W ballast.

Table 1.15   Recommended levels of illumination

Premises	Illumination Level (lumens per square metre or lux)
Residential	
Living room, kitchen, bedroom	150
Dining room	150
Study room	300
Corridor, toilet, staircase	50–100
Offices	
General	200–300
Drawing, architecture, typing	400–1500
Factories	
Assembly, rough and medium	150–250
Fine and very fine assembly	
of components, precious stone	
cutting, optics and watchmaking,	
machine shop	
Printing (general)	
Composing	500
Colour print inspection	800
Textile carding, spinning	150–250
Weaving coarse	350
Weaving fine	800
Wood working	200–800
Welding general precision	250–350
Inspection medium and fine	350–800
Very fine and minute	1500–2500
Schools	
Classrooms	250–350
Classrooms for chemistry, physics,	
needle works, arts	500

(*Contd...*)

Table 1.15 Recommended levels of illumination (*Contd.*)

Premises	Illumination level (*lumens per square metre or lux*)
Shops	
General	500–1000
Show windows	1000–2000
Hospitals	
General	150–200
Operation theatre	500–1000
Operation table	Over 5000
Gardens	5–30
Railway yards	5–20
Canteen's	150–250
Car parks	5–10
Traffic routes according to traffic density	50–100

Table 1.16   Specific gravity

Metal	Sp gr	Metal	Sp gr
Aluminium (cast)	2.56	Nickel	8.28
Aluminium (wrought)	2.67	Nickel hammered	8.67
Bismuth	9.90	Phosphor bronze	8.60
Brass, cast	8.10	Platinum, rolled	22.10
Brass, sheet, 75% Cu	8.45	Silver pure cast	10.47
Copper, cast	8.79	Silver, pure hammered	10.50
Copper, sheet	8.81	Steel	7.73
Copper, wire	8.91		to
Gold, pure cast	19.30		7.90
Gold, 22 caarats	17.50	Steel mean	7.85
Gun-metal 83% Cu	8.46	Tin (hammered)	7.39
Gun-metal 79% Cu	8.73	Tin (pure)	7.29
Iron, cast	6.90	Zinc cast	6.86
	to	Zinc rolled *Plastics*	7.19
	7.50	Plaskon (urea)	1.5
Iron, cast (mean)	7.22	Polystyrene	1.06
Iron wrought	7.70	Casein	1.33
Lead	11.40	Polyvinyl acetate	1.9
Mercury at 60 °F	13.60	Acrylate	1.49
Monel	8.87	Furfural phenol	1.35

Table 1.17   Melting point and specific heat of metals

Metal	Melting Point		Specific Heat
	°F	°C	
Aluminium	1210	658	.219
80 copper: 20 zinc	1850	1010	.092
Brass			
50 copper: 50 zinc	1700	927	–
Bronze	1675	912	–
Cast iron	2200	1186	.144
Copper	1981	1083	.0936
Lead	620	327	.0305
Monel	2480	1360	–
Nickel	2600	1452	.109
Steel	2500	1371	.116
Tin	446	230	.0553
Zinc	785	419	.0935

Table 1.18   Coefficient of linear expansion α per degree fahrenheit (°F)

Substance	α	Substance	α
Metals and alloys		**Stone and masonry**	
Aluminium wrought	.0000128	Ashlar masonry	.0000035
Brass	.0000104	Brick masonry	.0000031
Brass wire	.0000107	Cement, portland	.0000059
Bronze	.0000101	Concrete	.0000079
Copper	.0000093	Concrete masonry	.0000067
German silver	.0000102	Granite	.0000047
Gold	.0000083	Limestone	.0000044
Iron cast, grey	.0000059	Marble	.0000056
Iron wrought	.0000067	Plaster	.0000092
Iron, wire	.0000069	Rubble masonry	.0000035
Lead	.0000159	Sandstone	.0000061
Monel	.0000076	State	.0000058
Nickel	.0000070		

(Contd...)

Table 1.18 Coefficient of linear expansion α per degree fahrenheit (°F) (*Contd.*)

Substance	α	Substance	α
Platinum	.0000050	**Timber**	
Platinum-iridium 15% Ir.	.0000045	Fir Parallel	.0000021
Silver	.0000107	Maple to	.0000036
Steel, cast	.0000061	Oak fibre	.0000027
Steel, hard	.0000073	Pine	.0000030
Steel, medium	.0000067	Fir	.000032
Steel, soft	.0000061	Maple perpendi- cular	.0000027
Tin	.0000117	Oak to	.0000030
Zinc, rolled	.0000173	Pine fibre	.0000019
Miscellaneous solids		**Liquid substances**	*Vol. expansion*
Glass	.0000047	Alcohol	.00058
Graphite	.0000044	Acid, nitric	.00061
Gutta-percha	.0003322	Acid, sulphuric	.00035
Paraffin	.0001547	Mercury	.00010
Porcelain	.0000020	Oil, turpentine	.00050

5.   Thermometer and Hydrometer Scales

The number of degrees between freezing point and boiling point of water is $212 - 32 = 180$ degrees on the Fahrenheit scale, and 100 degrees on the Centigrade scale. The magnitude of the degrees F relative to degrees C is thus as 5 to 9.

$$\text{Temp (°C)} = 5/9 \ (°F - 32);$$

$$\text{Temp (°F)} = 9/5 \ (°C + 32).$$

The following formula enables degrees Baume to be converted into specific gravity.

$$\text{Sp gr} = \frac{140}{\text{Degs B} + 130};$$

$$\text{Degrees B} = \frac{140}{\text{Sp gr}} - 130$$

For liquids heavier than water: $\text{Sp gr} = \dfrac{146}{145 - \text{Degrees B}}.$

Table 1.19   Equivalent thermometer scales

°F	°C	°F	°C	°F	°C	°F	°C
2000	1093.3	212	100.0	100	37.7	50	10.0
1900	1037.8	200	93.3	98	36.6	49	9.4
1850	1010.0	190	87.7	96	35.5	48	8.8
1800	982.2	180	82.2	94	34.4	47	8.3
1750	954.4	170	76.6	92	33.3	46	7.7
1700	926.7	160	71.1	90	32.2	45	7.2
1650	898.9	158	70.0	88	31.1	44	6.6
1600	871.1	156	68.8	86	30.0	43	6.1
1550	843.3	154	67.7	84	28.8	42	5.5
1500	815.5	152	66.6	82	27.7	41	5.0
1450	787.8	150	65.5	80	26.6	40	4.4
1400	760.0	148	64.4	78	25.5	39	3.8
1350	732.2	146	63.3	76	24.4	38	3.3
1300	704.4	144	62.2	74	23.3	37	2.7
1250	676.7	142	61.1	72	22.2	36	2.2
1200	648.9	140	60.0	70	21.1	35	1.6
1150	621.1	138	58.8	69	20.5	34	1.1
1100	593.3	136	57.7	68	20.0	33	0.5
1050	565.5	134	56.6	67	19.4	32	0.0
1000	537.8	132	55.5	66	18.8	30	1.1
950	510.0	130	54.4	65	18.3	28	2.2
900	482.2	128	53.3	64	17.7	26	−3.3
850	454.4	126	52.2	63	17.2	24	−4.4
800	426.7	124	51.1	62	16.6	22	−5.5
750	398.9	122	50.0	61	16.1	20	−6.6
700	371.1	120	48.8	60	15.5	18	−7.7
650	343.3	118	47.7	59	15.0	16	−8.8
600	315.6	116	46.6	58	14.4	14	−10.0
550	287.8	114	45.5	57	13.8	12	−11.1
500	260.0	112	44.4	56	13.3	10	−12.2
450	232.2	110	43.3	55	12.7	8	−13.3
400	204.4	108	42.2	54	12.2	6	−14.4
350	176.7	106	41.1	53	11.6	4	−15.5
300	148.9	104	40.0	52	11.1	2	−16.6
250	121.1	102	38.8	51	10.5	0	−17.7

Chapter **2**

Algebra

1. Fundamental Properties (Real Numbers)

Commutative Law for Addition

$$a + b = b + a$$

Associative Law for Addition

$$(a + b) + c = a + (b + c)$$

Identity Law for Addition

$$a + 0 = a$$

Inverse Law for Addition

$$a + (-a) = (-a) + a = 0$$

Associative Law for Multiplication

$$a(bc) = (ab)c$$

Inverse Law for Multiplication

$$a\left(\frac{1}{a}\right) = \left(\frac{1}{a}\right)a = 1, \ a \neq 0$$

Identity Law for Multiplication

$$(a)(1) = (1)(a) = a$$

Commutative Law for Multiplication

$$ab = ba$$

18

Distributive Law

$$a(b + c) = ab + ac$$

Property of Zero $\quad a \cdot 0 = 0 \cdot a = 0$

Cancelation Law for Multiplication

If $a \cdot c = b \cdot c$ and $c \neq 0$ $\quad$ then $a = b$

Division by Zero is Not Defined

2. Exponents

For integers m and n

$$a^n a^m = a^{n+m}$$

$$a^n / a^m = a^{n-m}$$

$$(a^n)^m = a^{nm}$$

$$(ab)^m = a^m b^m$$

$$(a/b)^m = a^m / b^m$$

3. Fractional Exponents

$$a^{p/q} = (a^{1/q})^p$$

where $a^{1/q}$ is the positive qth root of a if $a > 0$ and the negative qth root of a if a is negative and q is odd. Accordingly, the five rules of exponents given above for integers are also valid if m and n are fractions, provided a and b are positive.

4. Irrational Exponents

If an exponent is irrational, *e.g.* $\sqrt{2}$, the quantity, such as $a^{\sqrt{2}}$ is the limit of the sequence $a^{1.4}$, $a^{1.41}$, $a^{1.414}$, ...

Operations with Zero

$$0^m = 0; \qquad a^0 = 1$$

5. Logarithms

If x, y and b are positive and $b \neq 1$

$$\log_b (xy) = \log_b x + \log_b y$$

$$\log_b(x/y) = \log_b x - \log_b y$$

$$\log_b x^n = n \log_b x$$

$$\log_b (1/x) = -\log_b x$$

$$\log_b b = 1$$

$$\log_b 1 = 0 \qquad \text{Note: } b^{\log_b^x} = x$$

Change of base $(a \neq 1)$

$$\log_b x = \log_a x. \log_b a \Rightarrow \log_a x = \frac{\log_b x}{\log_b a}$$

6. Factorials

The factorial of a positive integer n is the product of all the positive integers less than or equal to the integer n and is denoted by $n!$. Thus,

$$n! = 1 \cdot 2 \cdot 3 \cdot \ldots \cdot n.$$

Factorial 0 is defined as $0! = 1$.

Stirling's Approximation

$$\lim_{n \to \infty} (n / e)^n \sqrt{2\pi n} = n!$$

7. Factors and Expansion

$$(a + b)^2 = a^2 + 2ab + b^2$$

$$(a + b + c)^2 = a^2 + b^2 + c^2 + 2ab + 2bc + 2ca$$

$$(a + b)^2 = (a - b)^2 + 4ab$$

$$(a - b)^2 = a^2 - 2ab + b^2$$

$$(a - b)^2 = (a + b)^2 - 4ab$$

$$(a + b)^3 = a^3 + 3a^2b + 3ab^2 + b^3$$

$$(a - b)^3 = a^3 - 3a^2b + 3ab^2 - b^3$$

$$(a^2 - b^2) = (a - b)(a + b)$$

$$(a^3 - b^3) = (a - b)(a^2 + ab + b^2)$$

$$(a^3 + b^3) = (a + b)(a^2 - ab + b^2)$$

$$(a^3 + b^3 + c^3 - 3abc) = (a + b + c)(a^2 + b^2 + c^2 - ab - bc - ca)$$

8. Progression

An arithmetic progression is a sequence in which the difference between any term and the preceding term is a constant (d).

If the last term is denoted by l [$= a + (n - 1)d$], then the sum is

$$S = \frac{n}{2}(a+l)$$

nth term $T_n = a + (n - 1)d$

$$\Sigma n = 1 + 2 + 3 + \ldots \ldots n = \frac{n(n+1)}{2}$$

$$\Sigma n^2 = 1^2 + 2^2 + 3^2 + \ldots \ldots n^2 = \frac{n(n+1)(2n+1)}{6}$$

$$\Sigma n^3 = 1^3 + 2^3 + 3^3 + \ldots \ldots n^3 = \left[\frac{n(n+1)}{2}\right]^2$$

A geometric progression is a sequence in which the ratio of any term to the preceding term is a constant r. Thus for n terms, it is

$$S = a, ar, ar^2, \ldots, ar^{n-1}.$$

nth term $\qquad\qquad T_n = ar^{n-1}$

sum $\qquad\qquad S = \dfrac{a - ar^n}{1 - r} \qquad$ [if $r < 1$]

$\qquad\qquad\qquad = \dfrac{a(r^n - 1)}{r - 1} \qquad$ [if $r > 1$]

Sum to infinity $S_\infty = \dfrac{a}{1 - r}$

Geometric mean between two quantities $= \sqrt{ab}$

Three consecutive numbers in GP are: $\dfrac{a}{b}$, a, ar

Four consecutive numbers in GP are: $\dfrac{a}{r^3}$, $\dfrac{a}{r}$, ar, ar^3

9. Complex Numbers

A complex number is an ordered pair of real numbers (a, b).

Equality: $\qquad (a, b) = (c, d)$ if and only if $a = c$ and $b = d$

> **Addition:** $(a, b) + (c, d) = (a + c, b + d)$
>
> **Multiplication:** $(a, b)(c, d) = (ac - bd, ad + bc)$

The first element of (a, b) is called the *real* part; the second element is the *imaginary* part. An alternate notation for (a, b) is $a + bi$, where $i^2 = (-1, 0)$ and $i = (0, 1)$ or $0 + 1i$ is written for this complex number as a convenience. The conjugate of $a + bi$ is $a - bi$ and the product of a complex number and its conjugate is $a^2 + b^2$.

10. Polar Form

The complex number $x + iy$ may be represented by a plane vector with components x and y.

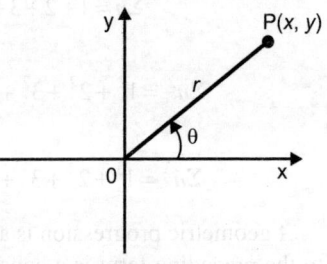

$$x + iy = r(\cos\theta + i\,\sin\theta)$$

Then, given two complex numbers (Fig. 2.1).
$z_1 = r_1(\cos\theta_1 + i\sin\theta_1)$ and
$z_2 = r_2(\cos\theta_2 + i\sin\theta_2)$,
the product and quotient are:

Fig. 2.1

product: $z_1 z_2 = r_1 r_2 [\cos(\theta_1 + \theta_2) + i\,\sin(\theta_1 + \theta_2)]$

quotient: $z_1/z_2 = (r_1/r_2)[\cos(\theta_1 - \theta_2) + i\,\sin(\theta_1 - \theta_2)]$

powers: $z^n = [r(\cos\theta + i\,\sin\theta)]^n = [r^n(\cos n\theta + i\,\sin n\theta)]$

roots: $z^{1/n} = [r(\cos\theta + i\,\sin\theta)]^{1/n}$

$$= r^{1/n}\left[\cos\frac{\theta + k\cdot 360}{n} + i\sin\frac{\theta + k\cdot 360}{n}\right];$$

$$k = 0, 1, 2,..., n - 1$$

11. Permutation

A permutation is an ordered arrangement (sequence) of all or part of a set of objects. The number of permutations of n objects taken r at a time,

$$p(n, r) = n(n - 1)(n - 2)...(n - r + 1)$$

$$= \frac{n!}{(n-r)!}, n! = n(n - 1)(n - 2)...3.2.1, \text{ if } n \text{ is a } +ve \text{ integer.}$$

12. Binomial Theorem

For positive integer n

$$(x + y)^n = \overset{n}{C_0}\, x^n + \overset{n}{C_1}\, x^{n-1} \cdot y + \overset{n}{C_2}\, x^{n-2} y^2 + \ldots \overset{n}{C_r}\, x^{n-r} y^r + \overset{n}{C_n}\, y^n$$

or $(x + y)^n = x^n + nx^{n-1}y + \dfrac{n(n-1)}{2!} x^{n-2} y^2$

$$+ \dfrac{n(n-1)(n-2)}{3!} x^{n-3} y^3 + \ldots + nxy^{n-1} + y^n$$

13. Combination

A combination is a selection of one or more objects from among a set of objects regardless of order. The number of combinations of n different objects taken r at a time

$$C(n,\, r) = \dfrac{P(n,\, r)}{r!} = \dfrac{n!}{r!(n-r)!}$$

14. Algebraic Equations

Quadratic

If $ax^2 + bx^2 + c = 0$, and $a \ne 0$, then roots are:

$$x = \dfrac{-b \pm \sqrt{b^2 - 4ac}}{2a}$$

If α and β are the roots of the quadratic equation $ax^2 + bx + c = 0$.

Then, $\alpha + \beta = -b/a$ and $\alpha\beta = \dfrac{c}{a}$

If the sum of the roots (S) and product of the roots (P) are known, then the equation is given by

Cubic $x^2 - Sx + P = 0$

To solve $x^3 + bx^2 + cx + d = 0$, let $x = y - b/3$. Then, the *reduced cubic* equation is obtained

$$y^3 + py + q = 0$$

where, $p = c - (1/3)b^2$ and $q = d - (1/3)bc + (2/27)b^3$.

Solutions of the original cubic in terms of the reduced cubic roots y_1, y_2, y_3:

$$x_1 = y_1 - (1/3)b \qquad\qquad x_2 = y_2 - (1/3)b$$

$$x_3 = y_3 - (1/3)b$$

The three roots of the reduced cubic are

$$y_1 = (A)^{1/3} + (B)^{1/3}$$

$$y_2 = W(A)^{1/3} + W^2(B)^{1/3}$$

$$y_3 = W^2(A)^{1/3} + W(B)^{1/3}$$

where

$$A = -\frac{1}{2}q + \sqrt{(1/27)p^3 + \frac{1}{4}q^2},$$

$$B = -\frac{1}{2}q + \sqrt{(1/27)p^3 + \frac{1}{4}q^2},$$

$$W = \frac{-1 + i\sqrt{3}}{2}, \ W^2 = \frac{-1 - i\sqrt{3}}{2}.$$

When $(1/27)p^3 + (1/4)q^2$ is negative, A is complex. In this case A should be expressed in trigonometric form: $A = r(\cos\theta + i\sin\theta)$, where θ is a first or second quadrant angle, as q is negative or positive. The three roots of the reduced cubic are

$$y_1 = 2(r)^{1/3} \cos(\theta/3)$$

$$y_2 = 2(r)^{1/3} \cos\left(\frac{\theta}{3} + 120°\right)$$

$$y_3 = 2(r)^{1/3} \cos\left(\frac{\theta}{3} + 240°\right)$$

Chapter 3

Plane Surfaces and Solids

Area and Volume

The following is a collection of common geometric figures. Area (A), volume (V), and other measurable features are indicated.

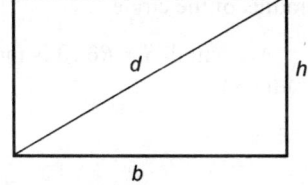

Fig. 3.1

1. Rectangle

Area of a Rectangle $A = bh$

Diagonal $d = \sqrt{b^2 + h^2}$

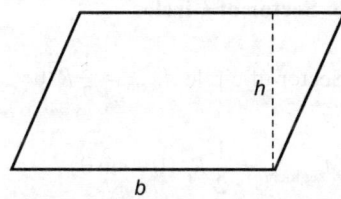

Fig. 3.2

2. Parallelogram

Area of parallelogram $A = bh$

3. Triangle

Area of triangle $A = \dfrac{1}{2}bh$

Area of the triangle

$$A = \sqrt{s(s-a)(s-b)(s-c)}$$

where, $s = \dfrac{a+b+c}{2}$ and a, b and c are the sides of the triangle.

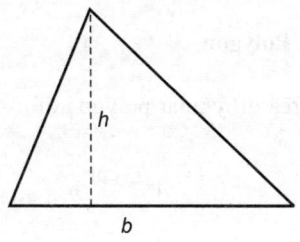

Fig. 3.3

25

4. Trapezoid

Area of trapezoid $A = \dfrac{1}{2}(a+b)h$

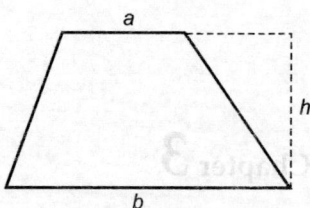

Fig. 3.4

5. Circle

Area of circle $A = \pi R^2$

Circumference $= 2\pi R$; where R is the radius of the circle

Arc length $S = R\theta$ (θ is measured in radians)

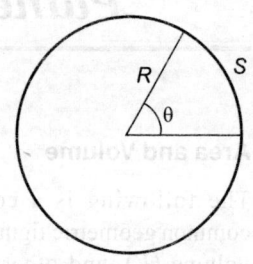

Fig. 3.5

6. Sector of Circle

Sector of circle $A_{\text{sector}} = \dfrac{1}{2}R^2\theta$;

$A_{\text{segment}} = \dfrac{1}{2}R^2(\theta - \sin\theta)$

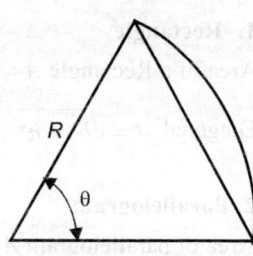

Fig. 3.6

7. Polygon

Area of regular polygon of n sides

$$A = \frac{n}{4}b^2 \cot\frac{\pi}{n}$$

$$R = \frac{b}{2}\operatorname{cosec}\frac{\pi}{n}$$

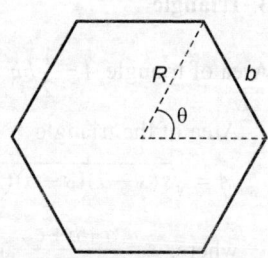

Fig. 3.7

8. Cylinder

Volume of a right circular cylinder,
$V = \pi R^2 h$

Lateral surface area $= 2\pi Rh$

Total surface area $= (2\pi Rh + 2\pi R^2)$

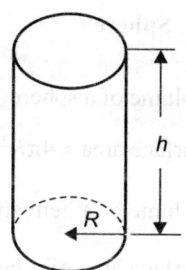

Fig. 3.8

9. Prism

Lateral surface of prism = perimeter of base × height

Volume of cylinder (or prism) with parallel bases $V = Ah$

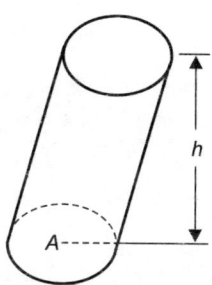

Fig. 3.9

10. Circular Cone

Volume of right circular cone

$$V = \frac{1}{3}\pi R^2 h;$$

Lateral surface area

$$= \pi Rl = \pi R\sqrt{R^2 + h^2}$$

Total surface area

$$= \pi R(l + R)$$

$$= \pi R\left(\sqrt{R^2 + h^2} + R\right)$$

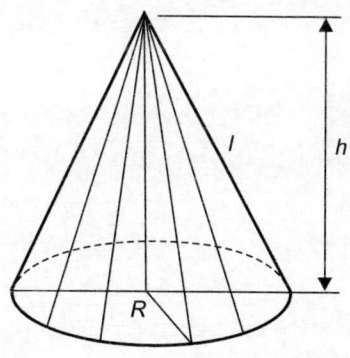

Fig. 3.10

11. Sphere

Volume of a sphere $V = \dfrac{4}{3}\pi R^3$

Surface area $= 4\pi R^2$

Volume of a hemisphere $= \dfrac{2}{3}\pi R^2$

Surface area of a hemisphere $= 3\pi R^2$

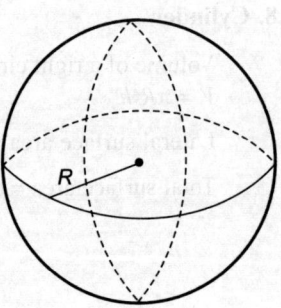

Fig. 3.11

Chapter 4

Trigonometry

1. Trigonometric Ratios

In a right angled triangle; if any acute angle be taken as θ then

$\sin \theta = P/H$, $\mathrm{cosec}\ \theta = H/P$

$\cos \theta = B/H$, $\sec \theta = H/B$

$\tan \theta = P/B$, $\cot \theta = B/P$

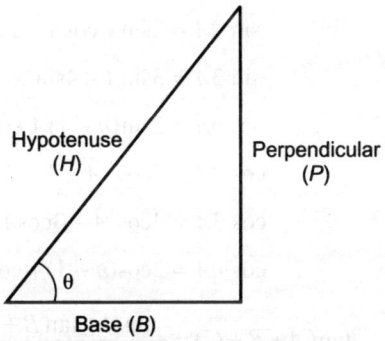

Fig. 4.1

2. Trigonometric Identities

$$\sin \theta = \frac{1}{\mathrm{cosec}\ \theta}; \qquad \cos \theta = \frac{1}{\sec \theta}$$

$$\tan \theta = \frac{1}{\cot \theta} = \frac{\sin \theta}{\cos \theta}; \quad \mathrm{cosec}\ \theta = \frac{1}{\sin \theta}$$

$$\sec \theta = \frac{1}{\cos \theta}; \qquad \cot \theta = \frac{1}{\tan \theta} = \frac{\cos \theta}{\sin \theta}$$

$$\sin^2\theta + \cos^2\theta = 1$$

$$1 + \tan^2\theta = \sec^2\theta$$

$$1 + \cot^2\theta = \mathrm{cosec}^2\theta$$

3. Trigonometric Ratios of Compound Angles

$$\sin(A + B) = \sin A \cos B + \cos A \sin B$$
$$\sin(A - B) = \sin A \cos B - \cos A \sin B$$
$$\cos(A + B) = \cos A \cos B - \sin A \sin B$$
$$\cos(A - B) = \cos A \cos B + \sin A \sin B$$

$$\tan(A+B) = \frac{\tan A + \tan B}{1 - \tan A \tan B}$$

$$\tan(A-B) = \frac{\tan A - \tan B}{1 + \tan A \tan B}$$

4. Multiple Angles

$$\sin 2A = 2\sin A \cos A = 2\tan A/(1 - \tan^2 A)$$
$$\sin 3A = 3\sin A - 4\sin^3 A$$
$$\sin nA = 2\sin(n - 1)A \cos A - \sin(n - 2)A$$
$$\cos 2A = 2\cos^2 A - 1 = 1 - 2\sin^2 A = 1 - \tan^2 A/(1 + \tan^2 A)$$
$$\cos 3A = 4\cos^3 A - 3\cos A$$
$$\cos nA = 2\cos(n - 1)A \cos A - \cos(n - 2)A$$

$$\tan(A+B+C) = \frac{\tan A + \tan B + \tan C - \tan A \tan B \tan C}{1 + \tan A \tan B - \tan B \tan C - \tan C \tan A}$$

$$\tan 2A = 2\tan A/(1 - \tan^2 A)$$

$$\tan 3A = \frac{3\tan A - \tan^3 A}{1 - 3\tan^2 A}$$

5. Sum and Product Formulae

$$\sin A + \sin B = 2\sin\frac{1}{2}(A+B)\cos\frac{1}{2}(A-B)$$

$$\sin A - \sin B = 2\cos\frac{1}{2}(A+B)\sin\frac{1}{2}(A-B)$$

$$\cos A + \cos B = 2\cos\frac{1}{2}(A+B)\cos\frac{1}{2}(A-B)$$

$$\cos A - \cos B = -2\sin\frac{1}{2}(A+B)\sin\frac{1}{2}(A-B)$$

$$\tan A \pm \tan B = \frac{\sin(A\pm B)}{\cos A \cos B}$$

$$\cot A \pm \cot B = \pm\frac{\sin(A\pm B)}{\sin A \sin B}$$

$$\sin A \sin B = \frac{1}{2}\cos(A-B) - \frac{1}{2}\cos(A+B)$$

$$\cos A \cos B = \frac{1}{2}\cos(A-B) + \frac{1}{2}\cos(A+B)$$

$$\sin A \cos B = \frac{1}{2}\sin(A+B) + \frac{1}{2}\sin(A-B)$$

$$\sin\frac{A}{2} = \pm\sqrt{\frac{1-\cos A}{2}}$$

$$\cos\frac{A}{2} = \pm\sqrt{\frac{1+\cos A}{2}}$$

$$\tan\frac{A}{2} = \frac{1-\cos A}{\sin A} = \frac{\sin A}{1+\cos A} = \pm\sqrt{\frac{1-\cos A}{1+\cos A}}$$

6. Submultiple Angles

$$\sin^2 A = \frac{1}{2}(1-\cos 2A) = \frac{2\tan A}{1+\tan^2 A}$$

$$\cos^2 A = \frac{1}{2}(1+\cos 2A)$$

$$\tan^2 A = \frac{1-\cos 2A}{1+\cos 2A}$$

$$\sin^3 A = \frac{1}{4}(3\sin A - \sin 3A) = \frac{1-\tan^2 A}{1+\tan^2 A}$$

$$\cos^3 A = \frac{1}{4}(\cos 3A + 3\cos A)$$

$$\sin ix = \frac{1}{2}i(e^x - e^{-x}) = i\sinh x$$

$$\cos ix = \frac{1}{2}(e^x + e^{-x}) = \cosh x$$

$$\tan ix = \frac{i(e^x - e^{-x})}{e^x + e^{-x}} = i \tanh x$$

$$e^{x+iy} = e^x(\cos y + i \sin y)$$

$$(\cos x \pm i \sin x)^n = \cos nx + i \sin nx$$

7.   Inverse Trigonometric Functions

The inverse trigonometric functions are multiple valued, and this should be taken into account in the use of the following formulae.

$$\sin^{-1} x = \cos^{-1} \sqrt{1 - x^2}$$

$$= \tan^{-1} \frac{x}{\sqrt{1-x^2}} = \cot^{-1} \frac{\sqrt{1-x^2}}{x}$$

$$= \sec^{-1} \frac{1}{\sqrt{1-x^2}} = \operatorname{cosec}^{-1} \frac{1}{x}$$

$$= -\sin^{-1}(-x)$$

$$\cos^{-1} x = \sin^{-1} \sqrt{1 - x^2}$$

$$= \tan^{-1} \frac{\sqrt{1-x^2}}{x} = \cot^{-1} \frac{x}{\sqrt{1-x^2}}$$

$$= \sec^{-1} \frac{1}{x} = \operatorname{cosec}^{-1} \frac{1}{\sqrt{1-x^2}}$$

$$= \pi - \cos^{-1}(-x)$$

$$\tan^{-1} x = \cot^{-1} \frac{1}{x}$$

$$= \sin^{-1} \frac{x}{\sqrt{1+x^2}} = \cos^{-1} \frac{1}{\sqrt{1+x^2}}$$

$$= \sec^{-1} \sqrt{1+x^2} = \operatorname{cosec}^{-1} \frac{\sqrt{1+x^2}}{x}$$

$$= -\tan^{-1}(-x)$$

8. Relation between the Sides and Angles of a Triangle

In any triangle (in a plane) with sides a, b and c and corresponding opposite angles A, B, C

$$a = b \cos C + c \cos B, \quad b = c \cos A + a \cos C$$

Sine formula $\qquad \dfrac{a}{\sin A} = \dfrac{b}{\sin B} = \dfrac{c}{\sin C}$

Cosine formula $\qquad a^2 = b^2 + c^2 - 2cb \cos A$

Tangent formula $\qquad \dfrac{a+b}{a-b} = \dfrac{\tan \frac{1}{2}(A+B)}{\tan \frac{1}{2}(A-B)}$

$$\sin \frac{1}{2} A = \sqrt{\frac{(s-b)(s-c)}{bc}}, \text{ where } s = \frac{1}{2}(a+b+c)$$

$$\cos \frac{1}{2} A = \sqrt{\frac{s(s-a)}{bc}}$$

$$\tan \frac{1}{2} A = \sqrt{\frac{(s-b)(s-c)}{s(s-a)}}$$

$$\text{Area of the triangle} = \frac{1}{2} bc \sin A$$

where, $\qquad \sin A = \dfrac{2}{bc} \sqrt{\{s(s-a)(s-b)(s-c)\}}$

$\therefore \quad$ Area of the triangle $= \sqrt{\{s(s-a)(s-b)(s-c)\}}$

Circum radius $R = \dfrac{a}{2 \sin A} = \dfrac{b}{2 \sin B}$

$$= \frac{c}{2 \sin C} = \frac{abc}{4A}$$

9. Trigonometric Functions

The trigonometric functions of $0°$, $30°$, $45°$, $60°$ and integer multiples of these are directly computed (Table 4.1).

Table 4.1 Values of trigonometric ratios with relative angles 0°–180°

Trigonometric ratios / Angles	0°	30°	45°	60°	90°	120°	135°	150°	180°
sin	0	$\dfrac{1}{2}$	$\dfrac{\sqrt{2}}{2}=\dfrac{1}{\sqrt{2}}$	$\dfrac{\sqrt{3}}{2}$	1	$\dfrac{\sqrt{3}}{2}$	$\dfrac{\sqrt{2}}{2}=\dfrac{1}{\sqrt{2}}$	$\dfrac{1}{2}$	0
cos	1	$\dfrac{\sqrt{3}}{2}$	$\dfrac{\sqrt{2}}{2}=\dfrac{1}{\sqrt{2}}$	$\dfrac{1}{2}$	0	$-\dfrac{1}{2}$	$-\dfrac{\sqrt{2}}{2}$	$-\dfrac{\sqrt{3}}{2}$	-1
tan	0	$\dfrac{\sqrt{3}}{3}=\dfrac{1}{\sqrt{3}}$	1	$\sqrt{3}$	∞	$-\sqrt{3}$	-1	$-\dfrac{\sqrt{3}}{3}$	0
cot	∞	$\sqrt{3}$	1	$\dfrac{\sqrt{3}}{3}=\dfrac{1}{\sqrt{3}}$	0	$-\dfrac{\sqrt{3}}{3}=-\dfrac{1}{\sqrt{3}}$	-1	$-\sqrt{3}$	∞
sec	1	$\dfrac{2\sqrt{3}}{3}=\dfrac{2}{\sqrt{3}}$	$\sqrt{2}$	2	∞	-2	$-\sqrt{2}$	$-\dfrac{2\sqrt{3}}{\sqrt{3}}$	-1
cosec	∞	2	$\sqrt{2}$	$\dfrac{2\sqrt{3}}{3}=\dfrac{2}{\sqrt{3}}$	1	$\dfrac{2\sqrt{3}}{3}=\dfrac{2}{\sqrt{3}}$	$\sqrt{2}$	2	∞

Chapter 5

Table of Derivatives

1. Derivatives

In the following explanations, a and n are constants, e is the base of the natural logarithm, and u an v denote functions of x.

    i. $\dfrac{d}{dx}(a) = 0$

    ii. $\dfrac{d}{dx}(x^n) = n \cdot x^{n-1}$

    iii. $\dfrac{d}{dx}(au) = a\dfrac{du}{dx}$

    iv. $\dfrac{d}{dx}(u+v) = \dfrac{du}{dx} + \dfrac{dv}{dx}$

    v. $\dfrac{d}{dx}(uv) = u\dfrac{dv}{dx} + v\dfrac{du}{dx}$

    vi. $\dfrac{d}{dx}\left(\dfrac{u}{v}\right) = \dfrac{v\,du/dx - u\,dv/dx}{v^2}$

    vii. $\dfrac{d}{dx}(u^n) = nu^{n-1}\dfrac{du}{dx}$

    viii. $\dfrac{d}{dx}e^u = e^u\dfrac{du}{dx}$

    ix. $\dfrac{d}{dx}a^u = (\log_e a)a^u\dfrac{du}{dx}$

x. $\dfrac{d}{dx}\log_e u = \left(\dfrac{1}{u}\right)\dfrac{du}{dx}$

xi. $\dfrac{d}{dx}\log_a u = (\log_a e)\left(\dfrac{1}{u}\right)\dfrac{du}{dx}$

xii. $\dfrac{d}{dx}u^v = vu^{v-1}\dfrac{du}{dx} + u^v (\log_e u)\dfrac{dv}{dx}$

xiii. $\dfrac{d}{dx}\sin u = \cos u \dfrac{du}{dx}$

xiv. $\dfrac{d}{dx}\cos u = -\sin u \dfrac{du}{dx}$

xv. $\dfrac{d}{dx}\tan u = \sec^2 u \dfrac{du}{dx}$

xvi. $\dfrac{d}{dx}\cot u = -\operatorname{cosec}^2 u \dfrac{du}{dx}$

xvii. $\dfrac{d}{dx}\sec u = \sec u \tan u \dfrac{du}{dx}$

xviii. $\dfrac{d}{dx}\operatorname{cosec} u = -\operatorname{cosec} u \cot u \dfrac{du}{dx}$

xix. $\dfrac{d}{dx}\sin^{-1} u = \dfrac{1}{\sqrt{1-u^2}}\dfrac{du}{dx}$ $\left(-\dfrac{1}{2}\pi \le \sin^{-1} u \le \dfrac{1}{2}\pi\right)$

xx. $\dfrac{d}{dx}\cos^{-1} u = \dfrac{-1}{\sqrt{1-u^2}}\dfrac{du}{dx}$ $(0 \le \cos^{-1} u \le \pi)$

xxi. $\dfrac{d}{dx}(\log x) = \dfrac{1}{x}$

xxii. $\dfrac{d}{dx}\tan^{-1} u = \dfrac{1}{1+u^2}\dfrac{du}{dx}$

xxiii. $\dfrac{d}{dx}\cot^{-1} u = \dfrac{-1}{1+u^2}\dfrac{du}{dx}$

xxiv. $\dfrac{d}{dx}\sec^{-1}u=\dfrac{1}{u\sqrt{u^2-1}}\dfrac{du}{dx}$

$$\left(-\pi\le\sec^{-1}u\le\dfrac{-1}{2}\pi;\ 0\le\sec^{-1}u\le\dfrac{1}{2}\pi\right)$$

xxv. $\dfrac{d}{dx}\operatorname{cosec}^{-1}u=\dfrac{-1}{u\sqrt{u^2-1}}\dfrac{du}{dx}$

$$\left(-\pi<\operatorname{cosec}^{-1}u\le\dfrac{-1}{2}\pi;\ 0\le\operatorname{cosec}^{-1}u\le\dfrac{1}{2}\pi\right)$$

xxvi. $\dfrac{d}{dx}\sinh u=\cosh u\dfrac{du}{dx}$

xxvii. $\dfrac{d}{dx}\cosh u=\sinh u\dfrac{du}{dx}$

xxviii. $\dfrac{d}{dx}\tanh u=\operatorname{sech}^2 u\dfrac{du}{dx}$

xxix. $\dfrac{d}{dx}\coth u=-\operatorname{cosech}^2 u\dfrac{du}{dx}$

xxx. $\dfrac{d}{dx}\operatorname{sech} u=-\operatorname{sech} u\tanh u\dfrac{du}{dx}$

xxxi. $\dfrac{d}{dx}\operatorname{cosech} u=-\operatorname{cosech} u\coth u\dfrac{du}{dx}$

xxxii. $\dfrac{d}{dx}\sinh^{-1}u=\dfrac{1}{\sqrt{u^2+1}}\dfrac{du}{dx}$

xxxiii. $\dfrac{d}{dx}\cosh^{-1}u=\dfrac{1}{\sqrt{u^2-1}}\dfrac{du}{dx}$

xxxiv. $\dfrac{d}{dx}\tanh^{-1}u=\dfrac{1}{1-u^2}\dfrac{du}{dx}$

xxxv. $\dfrac{d}{dx}\coth^{-1}u=\dfrac{-1}{u^2-1}\dfrac{du}{dx}$

xxxvi. $\dfrac{d}{dx}\operatorname{sech}^{-1}u=\dfrac{-1}{u\sqrt{1-u^2}}\dfrac{du}{dx}$

xxxvii. $\dfrac{d}{dx}\operatorname{cosech}^{-1}u=\dfrac{-1}{u\sqrt{u^2+1}}\dfrac{du}{dx}$

2.   Additional Relations with Derivatives

$$\frac{d}{dt}\int_a^t f(x)\,dx = f(t)$$

$$\frac{d}{dt}\int_t^a f(x)\,dx = -f(t)$$

If $x = f(y)$, then

$$\frac{dy}{dx} = \frac{1}{dx/dy}$$

3.   Chain Rule

If $y = f(u)$ and $u = g(x)$, then

$$\frac{dy}{dx} = \frac{dy}{du}\cdot\frac{du}{dx}$$

If $x = f(t)$ and $y = g(t)$, then

$$\frac{dy}{dx} = \frac{g'(t)}{f'(t)},$$

and

$$\frac{d^2 y}{dx^2} = \frac{f'(t)g''(t) - g'(t)f''(t)}{[f'(t)]^3}$$

Chapter 6

Table of Indefinite and Definite Integrals

1. **Standard Integrals:** Basic forms (all logarithms are to base e and C is a constant of integration)

   i. $\int dx = x + C$

   ii. $\int x^n dx = \dfrac{x^{n+1}}{n+1} + C$ $\hspace{3cm}$ $(n \neq -1)$

   iii. $\int \dfrac{dx}{x} = \log x + C$

   iv. $\int e^x dx = e^x + C$

   v. $\int a^x dx = \dfrac{a^x}{\log a} + C$

   vi. $\int \sin x\, dx = -\cos x + C$

   vii. $\int \cos x\, dx = \sin x + C$

   viii. $\int \tan x\, dx = -\log \cos x + C$

   ix. $\int \sec^2 x\, dx = \tan x + C$

   x. $\int \operatorname{cosec}^2 x\, dx = -\cot x + C$

xi. $\int \sec x \tan x \, dx = \sec x + C$

xii. $\int \sin^2 x \, dx = \dfrac{1}{2}x - \dfrac{1}{2}\sin x \cos x + C$

xiii. $\int \cos^2 x \, dx = \dfrac{1}{2}x + \dfrac{1}{2}\sin x \cos x + C$

xiv. $\int \log x \, dx = x \log x - x + C$

2. Integrals of the Form *ax + b* and Trigonometric Forms

i. $\int (ax+b)^m \, dx = \dfrac{(ax+b)^{m+1}}{a(m+1)} + C, \text{ where } (m \neq -1)$

ii. $\int x(ax+b)^m \, dx = \dfrac{(ax+b)^{m+2}}{a^2(m+2)} - \dfrac{b(ax+b)^{m+1}}{a^2(m+1)} + C, \ (m \neq -1, -2)$

iii. $\int \dfrac{dx}{ax+b} = \dfrac{1}{a}\log(ax+b) + C$

iv. $\int \dfrac{dx}{(ax+b)^2} = -\dfrac{1}{a(ax+b)} + C$

v. $\int (\sin ax) \, dx = -\dfrac{1}{a}\cos ax + C$

vi. $\int (\sin^2 ax) \, dx = -\dfrac{1}{2a}\cos ax \sin ax + \dfrac{1}{2}x = \dfrac{1}{2}x - \dfrac{1}{4a}\sin 2ax + C$

vii. $\int \sin(a+bx) \, dx = -\dfrac{1}{b}\cos(a+bx) + C$

viii. $\int \dfrac{dx}{1 \pm \sin ax} = \mp \dfrac{1}{a}\tan\left(\dfrac{\pi}{4} \mp \dfrac{ax}{2}\right) + C$

ix. $\int \dfrac{\sin ax}{1 \pm \sin ax} \, dx = \pm x + \dfrac{1}{a}\tan\left(\dfrac{\pi}{4} \mp \dfrac{ax}{2}\right) + C$

x. $\int (\cos ax) \, dx = \dfrac{1}{a}\sin ax + C$

xi. $\int (\cos^2 ax)\, dx = \dfrac{1}{2a}\sin ax \cos ax + \dfrac{1}{2}x = \dfrac{1}{2}x + \dfrac{1}{4a}\sin 2ax + C$

xii. $\int (\sin mx)(\sin nx)\, dx = \dfrac{\sin (m-n)x}{2(m-n)} - \dfrac{\sin (m+n)x}{2(m+n)} + C;\ (m^2 \neq n^2)$

xiii. $\int (\cos mx)(\cos nx)\, dx = \dfrac{\sin (m-n)x}{2(m-n)} + \dfrac{\sin (m+n)x}{2(m+n)} + C;$

$$(m^2 \neq n^2)$$

xiv. $\int (\sin ax)(\cos ax)\, dx = \dfrac{1}{2a}\sin^2 ax + C$

xv. $\int (\sin mx)(\cos nx)\, dx = -\dfrac{\cos (m-n)x}{2(m-n)} - \dfrac{\cos (m+n)x}{2(m+n)} + C;$

$$(m^2 \neq n^2)$$

3. Logarithmic and Exponential Forms

i. $\int (\log x)\, dx = x \log x - x + C$

ii. $\int x(\log x)\, dx = \dfrac{x^2}{2}\log x - \dfrac{x^2}{4} + C$

iii. $\int x^2 (\log x)\, dx = \dfrac{x^3}{3}\log x - \dfrac{x^3}{9} + C$

iv. $\int e^x dx = e^x + C$

v. $\int e^{-x} dx = -e^{-x} + C$

vi. $\int e^{ax} dx = \dfrac{e^{ax}}{a} + C$

vii. $\int x e^{ax} dx = \dfrac{e^{ax}}{a^2}(ax-1) + C$

viii. $\int \dfrac{dx}{1+e^x} = x - \log(1+e^x) = \log\dfrac{e^x}{1+e^x} + C$

4. Definite integrals

i. $\int_1^\infty \dfrac{dx}{x^m} = \dfrac{1}{m-1} + C \qquad (m > 1)$

ii. $\int_0^\infty \dfrac{dx}{(1+x)\sqrt{x}} = \pi$

iii. $\int_0^\infty \dfrac{a\,dx}{a^2 + x^2} = \dfrac{\pi}{2}, (a > 0)\,0$, or if $a = 0$; $-\dfrac{\pi}{2}$, if $a < 0$

iv. $\int_0^\infty e^{-ax} dx = \dfrac{1}{a} \; (a > 0)$

v. $\int_0^\infty \dfrac{e^{-ax} - e^{-bx}}{x} dx = \log \dfrac{b}{a} \; (a, b > 0)$

Chapter 7

Vector Analysis

1. Vectors

Equal Vector: Two vectors are said to be equal, if their lengths are equal and their directions are same.

Like Vectors: Vectors which have the same sense of directions.
Co-initial Vectors: Vectors which have the same initial point.
Collinear Vectors: Vectors which are parallel to the same line.
Unit Vector: A vector whose length is one unit.
Null Vector: A vector in which initial and terminal point coincide.

Given the set of mutually perpendicular unit vectors i, j and k (Fig. 7.1), then any vector in the space may be represented as $F = ai + bj + ck$, where a, b and c are *components*.

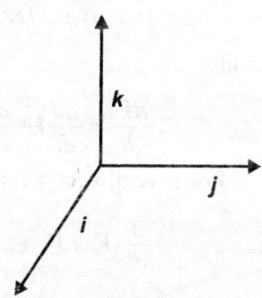

Fig. 7.1

a. *Magnitude of F*

$$|F| = \sqrt{(a^2 + b^2 + c^2)}$$

b. *Product by scalar p*

$$pF = pai + pbj + pck$$

c. *Sum of F_1 and F_2*

$$F_1 + F_2 = (a_1 + a_2)i + (b_1 + b_2)j + (c_1 + c_2)k$$

d. *Scalar product*

$$F_1 \cdot F_2 = a_1a_2 + b_1b_2 + c_1c_2$$

Thus, $i \cdot i = j \cdot j = k \cdot k = 1$ and $i \cdot j = j \cdot k = k \cdot i = 0$

43

Also,

$$F_1 \cdot F_2 = F_2 \cdot F_1$$

$$(F_1 + F_2) \cdot F_3 = F_1 \cdot F_3 + F_2 \cdot F_3$$

e. *Vector Product*

$$F_1 \times F_2 = \begin{vmatrix} i & j & k \\ a_1 & b_1 & c_1 \\ a_2 & b_2 & c_2 \end{vmatrix}$$

(Thus, $i \times i = j \times j = k \times k = 0$; $i \times j = k, j \times k = i$ and $k \times i = j$)

$$F_1 \times F_2 = -F_2 \times F_1$$

$$(F_1 + F_2) \times F_3 = F_1 \times F_3 + F_2 \times F_3$$

$$F_1 \times (F_2 + F_3) = F_1 \times F_2 + F_1 \times F_3$$

$$F_1 \times (F_2 \times F_3) = (F_1 \cdot F_3)F_2 - (F_1 \cdot F_2)F_3$$

$$F_1 \cdot (F_2 \times F_3) = (F_1 \times F_2) \cdot F_3$$

2.   Vector Differentiation

If V is a vector function of a scalar variable t, then

$$V = a(t)i + b(t)j + c(t)k$$

and

$$\frac{dV}{dt} = \frac{da}{dt}i + \frac{db}{dt}j + \frac{dc}{dt}k$$

For several vector functions $V_1, V_2, ..., V_n$

$$\frac{d}{dt}(V_1 + V_2 + ... + V_n) = \frac{dV_1}{dt} + \frac{dV_2}{dt} + ... + \frac{dV_n}{dt}$$

$$\frac{d}{dt}(V_1 \cdot V_2) = \frac{dV_1}{dt} \cdot V_2 + V_1 \cdot \frac{dV_2}{dt},$$

$$\frac{d}{dt}(V_1 \times V_2) = \frac{dV_1}{dt} \times V_2 + V_1 \times \frac{dV_2}{dt}.$$

For a scalar valued function $g(x, y, z)$

(Gradient) grad $g = \nabla g = \dfrac{\delta g}{\delta x}i + \dfrac{\delta g}{\delta y}j + \dfrac{\delta g}{\delta z}k$

For a vector valued function $V(a, b, c)$, where a, b, c, each is a function of x, y and z respectively.

(Divergence) $\text{div } V = \nabla \cdot V = \dfrac{\delta a}{\delta x} + \dfrac{\delta b}{\delta y} + \dfrac{\delta c}{\delta z}$

(Curl) $\text{curl } V = \nabla \times V = \begin{vmatrix} i & j & k \\ \dfrac{\delta}{\delta x} & \dfrac{\delta}{\delta y} & \dfrac{\delta}{\delta z} \\ a & b & c \end{vmatrix}$

Also,

$$\text{div grad } g = \nabla^2 g = \frac{\delta^2 g}{\delta x^2} + \frac{\delta^2 g}{\delta y^2} + \frac{\delta^2 g}{\delta z^2}$$

and $\quad$ curl grad $g = 0$ div curl $V = 0$;

$\quad\quad$ curl curl $V = \text{grad div } V - (i\nabla^2 a + j\nabla^2 b + k\nabla^2 c)$.

3. Divergence Theorem (Gauss)

Given a vector function F with continuous partial derivatives in a region R bounded by a closed surface S, then

$$\iiint_R \text{div } F \, dV = \iint_S n \cdot F \, dS,$$

where n is the (sectionally continuous) unit normal to S.

4. Stoke's Theorem

Given a vector function with continuous gradient over a surface S that consists of portions that are piecewise smooth and bounded by regular closed curves such as C, then

$$\iint_S n \cdot \text{curl } F \, dS = \oint_C F \cdot dr$$

5. Planer Motion in Polar Coordinates

Motion in a plane may be expressed with regard to polar coordinates (r, θ). Denoting the position vector by r and its magnitude by r, we have $r = rR\,(\theta)$, where R is the unit vector. Also, $dR/d\theta = P$, a unit vector perpendicular to R.

The velocity and acceleration are then

$$v = \frac{dr}{dt}R + r\frac{d\theta}{dt}P;$$

$$a = \left[\frac{d^2r}{dt^2} - r\left(\frac{d\theta}{dt}\right)^2\right]R + \left[r\frac{d^2\theta}{dt^2} + 2\frac{dr}{dt}\frac{d\theta}{dt}\right]P.$$

Note that the component of acceleration in P direction (transverse component) may also be written as

$$a = \frac{1}{r}\frac{d}{dt}\left(r^2\frac{d\theta}{dt}\right)$$

so that in purely radial motion it is zero and

$$r^2\frac{d\theta}{dt} = C \text{ (constant)}$$

which means that the position vector sweeps out area at a constant rate.

Chapter 8

Special Functions

1. Hyperbolic Functions

$$\sinh x = \frac{e^x - e^{-x}}{2}$$

$$\operatorname{cosech} x = \frac{1}{\sinh x}$$

$$\cosh x = \frac{e^x + e^{-x}}{2}$$

$$\operatorname{sech} x = \frac{1}{\cosh x}$$

$$\tanh x = \frac{e^x - e^{-x}}{e^x + e^{-x}}$$

$$\coth x = \frac{1}{\tanh x}$$

$$\sinh(-x) = -\sinh x$$

$$\coth x(-x) = -\coth x$$

$$\cosh(-x) = \cosh x$$

$$\operatorname{sech}(-x) = \operatorname{sech} x$$

$$\tanh(-x) = -\tanh x$$

$$\operatorname{cosech}(-x) = -\operatorname{cosech} x$$

$$\tanh x = \frac{\sinh x}{\cosh x}$$

$$\coth x = \frac{\cosh x}{\sin x}$$

$$\cosh^2 x - \sinh^2 x = 1$$

$$\cosh^2 x = \frac{1}{2}(\cosh 2x + 1)$$

$$\sinh^2 x = \frac{1}{2}(\cosh 2x - 1)$$

$$\coth^2 x - \operatorname{cosech}^2 x = 1$$

$$\operatorname{cosech}^2 x - \operatorname{sech}^2 x = \cosh^2 x \cdot \operatorname{sech}^2 x$$

$$\tanh^2 x + \operatorname{sech}^2 x = 1$$

$$\sinh(x + y) = \sinh x \cosh y + \cosh x \sinh y$$

47

$\cosh(x + y) = \cosh x \cosh y + \sinh x \sinh y$

$\sinh(x - y) = \sinh x \cosh y - \cosh x \sinh y$

$\cosh(x - y) = \cosh x \cosh y - \sinh x \sinh y$

$$\tanh(x + y) = \frac{\tanh x + \tanh y}{1 + \tanh x \tanh y}$$

$$\tanh(x - y) = \frac{\tanh x - \tanh y}{1 - \tanh x \tanh y}$$

2.   Gamma Function (Generalized Factorial Function)

The gamma function, denoted by $\Gamma(x)$, is defined by

$$\Gamma(x) = \int_0^\infty e^{-t} t^{x-1} dt \qquad (x > 0)$$

Properties

$$\Gamma(x + 1) = x\Gamma(x), \qquad\qquad (x > 0)$$

$$\Gamma(1) = 1$$

$$\Gamma(n + 1) = n\Gamma(n) = n! \qquad\qquad (n = 1, 2, 3, ...)$$

$$\Gamma(x)\Gamma(1 - x) = \pi/\sin \pi x$$

$$\Gamma\left(\frac{1}{2}\right) = \sqrt{\pi}$$

$$2^{2x-1}\Gamma(x)\Gamma\left(x + \frac{1}{2}\right) = \sqrt{\pi}\,\Gamma(2x)$$

3.   Laplace Transforms

The Laplace transform of the function $f(t)$, denoted by $F(s)$ or $\mathcal{L}\{f(t)\}$ is defined as

$$F(s) = \int_0^\infty f(t) e^{-st} dt$$

A sufficient condition for the existence of $F(s)$ is that $f(t)$ be of exponential order as $t \to \infty$ and that it is sectionally continuous over every finite interval in the range $t \geq 0$. The Laplace transform of $g(t)$ is denoted by $\mathcal{L}\{g(t)\}$ or $G(s)$.

Table 8.1 Laplace Transforms

$f(t)$	$F(s) = \int_0^\infty f(t)e^{-st}\,dt$
$af(t) + bg(t)$	$aF(s) + bG(s)$
$f'(t)$	$sF(s) - f(0)$
$f''(t)$	$s^2 F(s) - sf(0) - f'(0)$
$f^{(n)}(t)$	$s^n F(s) - s^{n-1} f(0) - s^{n-2} f'(0) - \ldots - f^{(n-1)}(0)$
$tf(t)$	$-F'(s)$
$t^n f(t)$	$(-1)^n F^{(n)}(s)$
$e^{at} f(t)$	$F(s - a)$
$\int_0^t f(t - \beta) \cdot g(\beta)\,d\beta$	$F(s) \cdot G(s)$
$f(t - a)$	$e^{-as} F(s)$
$f\left(\dfrac{t}{a}\right)$	$aF(as)$
$\int_0^t g(\beta)\,d\beta$	$\dfrac{1}{s} G(s)$
$f(t - c)\delta(t - c)$	$e^{-cs} F(s),\ c > 0$

where

$\delta(t - c) = 0$ if $0 \le t \le c$

$\quad\quad = 1$ if $t \ge c$

| $f(t) = f(t + \omega)$ (periodic) | $\dfrac{\displaystyle\int_0^\omega e^{-s\tau} f(\tau)\,dt}{1 - e^{-s\omega}}$ |

Table 8.2 Laplace Transforms of Functions

$f(t)$	$F(s)$	
1	$1/s$	
t	$1/s^2$	
$\dfrac{t^{n-1}}{(n-1)!}$	$1/s^n$	$(n=1,2,3,...)$
$\sqrt{t}$	$\dfrac{1}{2s}\sqrt{\dfrac{\pi}{s}}$	
$\dfrac{1}{\sqrt{t}}$	$\sqrt{\dfrac{\pi}{s}}$	
e^{at}	$\dfrac{1}{s-a}$	
te^{at}	$\dfrac{1}{(s-a)^2}$	
$\dfrac{t^{n-1}e^{at}}{(n-1)!}$	$\dfrac{1}{(s-a)^n}$	$(n=1,2,3,...)$
$\dfrac{t^x}{\Gamma(x+1)}$	$\dfrac{1}{s^{x+1}}$	$(x>-1)$
$\sin at$	$\dfrac{a}{s^2+a^2}$	
$\cos at$	$\dfrac{s}{s^2+a^2}$	
$\sinh at$	$\dfrac{a}{s^2-a^2}$	
$\cosh at$	$\dfrac{s}{s^2-a^2}$	
$e^{at}-e^{bt}$	$\dfrac{a-b}{(s-a)(s-b)},$	$(a\neq b)$
$ae^{at}-be^{bt}$	$\dfrac{s(a-b)}{(s-a)(s-b)},$	$(a\neq b)$

(Contd...)

Table 8.2 Laplace Transforms of Functions (*Contd.*)

$f(t)$	$F(s)$
$t \sin at$	$\dfrac{2as}{\left(s^2 + a^2\right)^2}$
$t \cos at$	$\dfrac{s^2 - a^2}{\left(s^2 + a^2\right)^2}$
$e^{at} \sin bt$	$\dfrac{b}{(s - a)^2 + b^2}$
$e^{at} \cos bt$	$\dfrac{s - a}{(s - a)^2 + b^2}$
$\dfrac{\sin at}{t}$	$\text{Arctan} \dfrac{a}{s}$
$\dfrac{\sinh at}{t}$	$\dfrac{1}{2} \log_e \left(\dfrac{s + a}{s - a}\right)$

4. z-Transform

For the real-valued sequence $\{f(k)\}$ and complex variable z, the z-transform, $F(z) = Z\{f(k)\}$ is defined by

$$Z\{f(k)\} = F(z) = \sum_{k=0}^{\infty} f(k) z^{-k}$$

For example, the sequence $f(k) = 1$, $k = 0, 1, 2$, has the z-transform

$$F(z) = 1 + z^{-1} + z^{-2} + z^{-3} + \ldots + z^{-k} + \ldots$$

5. z-Transform and the Laplace Transform

For the function $U(t)$, the output of the ideal sampler $U^*(t)$ is a set of values $U(kT)$, $k = 0, 1, 2, \ldots$, that is,

$$U^*(t) = \sum_{k=0}^{\infty} U(t) \delta(t - kT)$$

The Laplace transform of the output is

$$\mathcal{L}\{U^*(t)\} = \int_0^{\infty} e^{-st} U^*(t)\, dt = \int_0^{\infty} e^{-st} \sum_{k=0}^{\infty} U(t) \delta(t - kT)\, dt$$

$$= \sum_{k=0}^{\infty} e^{-skT} U(kT)$$

Defining $z = e^{sT}$ gives

$$\mathcal{L}\left\{U^*(t)\right\} = \sum_{k=0}^{\infty} U(kT)z^{-k}$$

which is the z-transform of the sampled signal $U(kT)$.

Properties

(i) *Linearity*: $Z\{af_1(k) + bf_2(k)\} = aZ\{f_1(k)\} + bZ\{f_2(k)\}$
$$= aF_1(z) + bF_2(z)$$
- *Right-shifting property*: $Z\{f(k - n)\} = z^{-n}F(z)$
- *Left-shifting property*: $Z\{f(k + n)\}$

$$= z^n F(z) - \sum_{k=0}^{n-1} f(k)z^{n-k}$$

(ii) *Time scaling*: $Z\{a^k f(k)\} = F(z/a)$

(iii) *Multiplication by k*: $Z\{kf(k)\} = -zdF(z)/dz$

(iv) *Initial value*: $f(0) = \lim_{z \to \infty}(1 - z^{-1})F(z) = F(\infty)$

(v) *Final value*: $\lim_{k \to \infty} f(k) = \lim_{z \to 1}(1 - z^{-1})F(z)$

(vi) *Convolution*: $Z\{f_1(k)\, f_2(k)\} = F_1(z)F_2(z)$

Table 8.3 z-Transforms of Sampled Functions

$f(k)$	$Z\{f(KT)\} = F(z)$
1 at k; else 0	z^{-k}
1	$\dfrac{z}{z-1}$
kT	$\dfrac{Tz}{(z-1)^2}$
$(kT)^2$	$\dfrac{T^2 z(z+1)}{(z-1)^3}$

(Contd...)

Table 8.3 *z*-Transforms of Sampled Functions (*Contd.*)

$f(k)$	$Z\{f(KT)\} = F(z)$
$\sin \omega kT$	$\dfrac{z \sin \omega T}{z^2 - 2z \cos \omega T + 1}$
$\cos \omega T$	$\dfrac{z(z - \cos \omega T)}{z^2 - 2z \cos \omega T + 1}$
e^{-akT}	$\dfrac{z}{z - e^{-aT}}$
kTe^{-akT}	$\dfrac{zTe^{-aT}}{(z - e^{-aT})^2}$
$(kT)^2 e^{-akT}$	$\dfrac{T^2 e^{-aT} z(z + e^{-aT})}{(z - e^{-aT})^3}$
$e^{-akT} \sin \omega kT$	$\dfrac{ze^{-aT} \sin \omega T}{z^2 - 2ze^{-aT} \cos \omega T + e^{-2aT}}$
$e^{-akT} \cos \omega kT$	$\dfrac{z(z - e^{-aT} \cos \omega T)}{z^2 - 2ze^{-aT} \cos \omega T + e^{-2aT}}$
$a^k \sin \omega kT$	$\dfrac{az \sin \omega T}{z^2 - 2az \cos \omega T + a^2}$
$a^k \cos \omega kT$	$\dfrac{z(z - a \cos \omega T)}{z^2 - 2az \cos \omega T + a^2}$

6. Fourier Series

The periodic function $f(t)$, with period 2π may be represented by the trigonometric series

$$a_0 + \sum_{1}^{\infty} (a_n \cos nt + b_n \sin nt),$$

where the coefficients are determined from

$$a_0 = \frac{1}{2\pi} \int_{-\pi}^{\pi} f(t)\, dt$$

$$a_n = \frac{1}{\pi} \int_{-\pi}^{\pi} f(t) \cos nt\, dt$$

$$b_n = \frac{1}{\pi} \int_{-\pi}^{\pi} f(t) \sin nt \, dt \quad (n = 1, 2, 3, ...)$$

Such a trigonometric series is called the *Fourier series* corresponding to *f(t) and the coefficients are termed Fourier coefficients of f(t)*. If the function is piece wise continuous in the interval $-\pi \le t \le \pi$, and has left and right-hand derivatives at each point in that interval, then the series is convergent with sum $f(t)$ except at points t_i at which $f(t)$ is discontinuous. At such points of discontinuity, the sum of the series is the arithmetic mean of the right and left-hand limits of $f(t)$ at t_i. The integrals in the formulas for the Fourier coefficients can have limits of integration that span a length of 2π, for example, 0 to 2π (because of the periodicity of the integrands).

7. Functions with Period Other Than 2π

If $f(t)$ has period P, the Fourier series is

$$f(t) \sim a_0 + \sum_{1}^{\infty} \left(a_n \cos \frac{2\pi n}{P} t + b_n \sin \frac{2\pi n}{P} t \right),$$

where

$$a_0 = \frac{1}{P} \int_{-P/2}^{P/2} f(t) \, dt$$

$$a_n = \frac{2}{P} \int_{-P/2}^{P/2} f(t) \cos \frac{2\pi n}{P} t \, dt$$

$$b_n = \frac{2}{P} \int_{-P/2}^{P/2} f(t) \sin \frac{2\pi n}{P} t \, dt.$$

The interval of integration in these formulae may be replaced by an interval of length P, for example 0 to P. Square wave is shown in Fig. 8.1.

$$f(t) \sim \frac{a}{2} + \frac{2a}{\pi} \left(\cos \frac{2\pi t}{P} - \frac{1}{3} \cos \frac{6\pi t}{P} + \frac{1}{5} \cos \frac{10\pi t}{P} + ... \right).$$

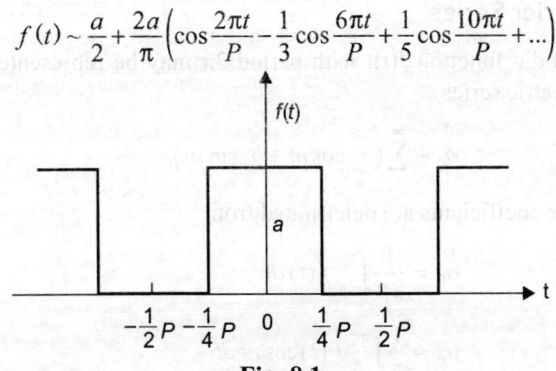

Fig. 8.1

Figure 8.2 shows the sawtooth wave.

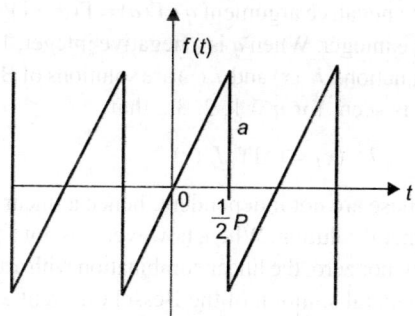

Fig. 8.2

$$f(t) \sim \frac{2a}{2}\left(\sin\frac{2\pi t}{P} - \frac{1}{2}\sin\frac{4\pi t}{P} + \frac{1}{3}\cos\frac{6\pi t}{P} + ...\right).$$

Half wave rectifier is shown in Fig. 8.3.

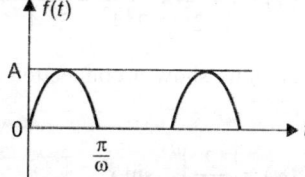

Fig. 8.3

$$f(t) \sim \frac{A}{\pi} + \frac{A}{2}\sin\omega t - \frac{2A}{\pi}\left(\frac{1}{(1)(3)}\cos 2\omega t + \frac{1}{(3)(5)}\cos 4\omega t + ...\right)$$

8. Bessel Functions

Bessel functions, also called *cylindrical functions*, arise in many physical problems as solutions of the differential equation

$$x^2 y'' + xy' + (x^2 - n^2)y = 0$$

which is known as Bessel's equation. Certain solutions of the above, known as *Bessel functions of the first kind of order n*, are given by

$$J_n(x) = \sum_{k=0}^{\infty} \frac{(-1)^k}{k!\,\Gamma(n+k+1)}\left(\frac{x}{2}\right)^{n+2k}$$

$$J_{-n}(x) = \sum_{k=0}^{\infty} \frac{(-1)^k}{k!\,\Gamma(-n+k+1)}\left(\frac{x}{2}\right)^{-n+2k}$$

In the above equations, it is noteworthy that the gamma function must be defined for the negative argument $q : \Gamma(q) = \Gamma(q + 1)/q$, provided that q is not a negative integer. When q is a negative integer, $1/\Gamma(q)$ is defined to be zero. The functions $J_{-n}(x)$ and $J_n(x)$ are solutions of Bessel's equation for all real n. It is seen, for $n = 1, 2, 3,...$that

$$J_{-n}(x) = (-1)^n J_n(x)$$

and, therefore, these are not independent; hence a linear combination of these, is not a general solution. When, however, n is not a positive integer, a negative integer, nor zero, the linear combination with arbitrary constants c_1 and c_2 is the general solution of the Bessel differential equation.

$$y = c_1 J_n(x) + c_2 J_{-n}(x)$$

The zero order function is especially important as it arises in the solution of the heat equation (for a "long" cylinder):

$$J_0(x) = 1 - \frac{x^2}{2^2} + \frac{x^4}{2^2 4^2} - \frac{x^6}{2^2\, 4^2\, 6^2} + ...$$

while the following relations show a connection to the trigonometric functions:

$$J_{\frac{1}{2}}(x) = \left[\frac{2}{\pi x}\right]^{1/2} \sin x$$

$$J_{-\frac{1}{2}}(x) = \left[\frac{2}{\pi x}\right]^{1/2} \cos x$$

The following recursion formula gives $J_{n+1}(x)$ for any order in terms of lower order functions:

$$\frac{2n}{x} J_n(x) = J_{n-1}(x) + J_{n+1}(x)$$

9. Legendre Polynomials

If Laplace's equation, $\nabla^2 V = 0$ is expressed in spherical coordinates, it is

$$r^2 \sin\theta \frac{\delta^2 V}{\delta r^2} + 2r\sin\theta \frac{\delta V}{\delta r} + \sin\theta \frac{\delta^2 V}{\delta\theta^2} + \cos\theta \frac{\delta V}{\delta\theta} + \frac{1}{\sin\theta} \frac{\delta^2 V}{\delta\phi^2} = 0$$

and any of its solutions, $V(r, \theta, \phi)$, are known as *spherical harmonics*.

The solution as a product

$$V(r, \theta, \phi) = R(r)\Theta(\theta)$$

which is independent of ϕ, leads to

$$\sin^2\theta\,\Theta'' + \sin\theta\,\cos\theta\,\Theta' + [n(n+1)\sin^2\theta]\Theta = 0$$

Rearrangement and substitution of $x = \cos\theta$ leads to

$$(1-x^2)\frac{d^2\Theta}{dx^2} - 2x\frac{d\Theta}{dx} + n(n+1)\Theta = 0$$

known as *Legendre's equation*. Important special cases are those in which n is zero or a positive integer, and, for such cases, Legendre's equation is satisfied by polynomials called *Legendre polynomials*, $P_n(x)$. A short list of Legendre polynomials, expressed in terms of x and $\cos\theta$, is given below.

$$P_n(x) = \sum_{j=0}^{L} \frac{(-1)^j(2n-2j)!}{2^n\,j!(n-j)!(n-2j)!}x^{n-2j}$$

where $L = n/2$ if n is even and

$L = (n-1)/2$ if n is odd.

Some are given below:

$P_0(x) = 1;$ $P_0(\cos\theta) = 1$

$P_1(x) = x;$ $P_1(\cos\theta) = \cos\theta$

$P_2(x) = \dfrac{1}{2}(3x^2 - 1);$ $P_2(\cos\theta) = \dfrac{1}{4}(3\cos2\theta + 1)$

$P_3(x) = \dfrac{1}{2}(5x^3 - 3x);$ $P_3(\cos\theta) = \dfrac{1}{8}(5\cos3\theta + 3\cos\theta)$

$P_4(x) = \dfrac{1}{8}(35x^4 - 30x^2 + 3);$ $P_4(\cos\theta) = \dfrac{1}{64}(35\cos4\theta + 20\cos2\theta + 9)$

$P_5(x) = \dfrac{1}{8}(63x^5 - 70x^3 + 15x);$

Additional Lengendre polynomials may be determined from the *recursion formula*,

$$(n+1)P_{n+1}(x) - (2n+1)\,xP_n(x) + nP_{n-1}(x) = 0 \quad (n = 1, 2, ...),$$

and it can also derived from the Rodrigues formula,

$$P_n(x) = \frac{1}{2^n \, n!} \frac{d^n}{dx^n} (x^2 - 1)^n$$

10. Laguerre Polynomials

Laguerre polynomials, denoted by $L_n(x)$, are solutions of the differential equation.

$$xy'' + (1 - x)y' + ny = 0$$

and are given by

$$L_n(x) = \sum_{j=0}^{n} \frac{(-1)^j}{j!} C_{(n,j)} x^j \quad (n = 0, 1, 2,...), \text{ thus}$$

$$L_0(x) = 1$$

$$L_1(x) = 1 - x$$

$$L_2(x) = 1 - 2x + \frac{1}{2}x^2$$

$$L_3(x) = 1 - 3x + \frac{3}{2}x^2 - \frac{1}{6}x^3$$

Additional Laguerre polynomials may be obtained from the recursion formula,

$$(n + 1)L_{n+1}(x) - (2n + 1 - x)L_n(x) + nL_{n-1}(x) = 0$$

11. Hermite Polynomials

The Hermite polynomials, denoted $H_n(x)$ are given by

$$H_0 = 1, \, H_n(x) = (-1)^n e^{x^2} \frac{d^n e^{-x^2}}{dx^n}, \quad\quad (n = 1, 2, ...)$$

and are solutions of the differential equation

$$y''n - 2xy' + 2ny = 0 \quad\quad (n = 0, 1, 2, ...)$$

The first few Hermite polynomials are

$$H_0 = 1 \quad\quad\quad H_1(x) = 2x$$

$$H_2(x) = 4x^2 - 2 \quad\quad\quad H_3(x) = 8x^3 - 12x$$

$$H_4(x) = 16x^4 - 48x^2 + 12$$

Additional Hermite polynomials may be obtained from the relation

$$H_{n+1}(x) = 2xH_n(x) - H_n'(x),$$

where prime denote differentiation with respect to x.

12. Orthogonality

A set of functions $\{f_n(x)\}$ $(n = 1, 2, ...)$ is orthogonal in an interval (a, b) with respect to a given weight function $w(x)$, if

$$\int_a^b w(x) f_m(x) f_n(x) dx = 0; \quad \text{when } m \neq n$$

The following polynomials are orthogonal on the given interval for the given $w(x)$.

Legendre polynomials $\qquad P_n(x) \qquad w(x) = 1$

$$a = -1, b = 1$$

Leguerre polynomials $\qquad L_n(x) \qquad w(x) = \exp(-x)$

$$a = 0, b = \infty$$

Hermite polynomials $\qquad H_n(x) \qquad w(x) = \exp(-x^2)$

$$a = -\infty, b = \infty$$

The *Bessel functions of order* n, $J_n(\lambda_1 x)$, $J_n(\lambda_2 x)$,..., are orthogonal with respect to $w(x) = x$ over the interval $(0, c)$ provided that the values of λ_i are positive roots of $J_n(\lambda c) = 0$;

$$\int_0^c x J_n(\lambda_j x) J_n(\lambda_k x) dx = 0 \quad (j \neq k)$$

where n is fixed and $n \geq 0$.

Chapter 9

Statistics

1. Arithmetic Mean

$$\mu = \frac{\Sigma X_i}{N},$$

where X_i is a measurement in the population and N is the total number of X_i in the population. For a *sample* of size n the sample mean, denoted $\bar{X}$.

$$\bar{X} = \frac{\Sigma X_i}{n}$$

2. Median

The median is the middle measurement when an odd number (n) of measurements is arranged in order; if n is even, it is the midpoint between the two middle measurements.

3. Mode

It is the most frequently occurring measurement in a set.

4. Geometric Mean

$$\text{Geometric mean} = \sqrt[n]{X_1 X_2 \ldots X_n}$$

5. Harmonic Mean

The harmonic mean H of n numbers $X_1, X_2, \ldots, X_n$.

$$H = \frac{n}{\Sigma(1/X_i)}$$

6. Variance

The mean of the sum of squares of deviations from the mean (μ) is the population variance, denoted by σ^2.

$$\sigma^2 = \Sigma(X_i - \mu)^2/N.$$

The sample variance s^2 for a sample size n

$$= \Sigma(X_i - \bar{X})^2/(n-1).$$

A simpler computational form can be given as below.

$$s^2 = \frac{\Sigma X_i^2 - (\Sigma X_i)^2/n}{n-1}$$

7. Standard Deviation

The positive square root of the population variance is the standard deviation. For a population,

$$\sigma = \left[\frac{\Sigma X_i^2 - (\Sigma X_i)^2/N}{N}\right]^{1/2};$$

for a sample,

$$s = \left[\frac{\Sigma X_i^2 - (\Sigma X_i)^2/n}{n-1}\right]^{1/2};$$

8. Coefficient of Variation

$$V = s/\bar{X}.$$

9. Moments

The rth moment about the mean of a distribution is denoted by μ_r and given by:

$$\mu_r = \frac{1}{N}\Sigma f(X - \bar{X})^r$$

10. Skewness

It measures the degree of asymmetry.

$$\text{Skewnesss} = \frac{\text{Mean} - \text{mode}}{\text{Standard deviation}}$$

11. Kurtosis

Kurtosis measures the degree of peakedness of a distribution.

12. Probability

For the sample space U, with subsets A of U (called *events*), we consider the probability measure of an event A to be a real-valued function p defined over all subsets of U such that: $0 \leq p(A) \leq 1$

Mathematically, $p(U) = 1$ and $p(\Phi) = 0$

If A_1 and A_2 are subsets of U

$$p(A_1 \cup A_2) = p(A_1) + p(A_2) - p(A_1 \cap A_2)$$

Two events A_1 and A_2 are called *mutually exclusive* if and only if $A_1 \cap A_2 = \phi$ (null set). These events are said to be independent if and only if $p(A_1 \cap A_2) = p(A_1)p(A_2)$.

13. Conditional Probability and Bayes' Rule

The probability of an event A, given that an event B has occurred is called the *conditional probability* and is denoted by $p(A/B)$. Further

$$p(A/B) = \frac{P(A \cap B)}{p(B)}$$

Bayes' rule permits a calculation of *a posteriori* probability from a given *priori* probabilities and is stated as below.

If A_1, A_2,..., A_n are n mutually exclusive events, and $p(A_1) + p(A_2) + ... + P(A_n) = 1$, and B is any event such that $p(B)$ is not 0, then the conditional probability $p(A_i/B)$ for any one of the events A_i *given that B has occurred* is

$$p(A_i/B) = \frac{p(A_i)p(B/A_i)}{p(A_1)p(B/A_1) + p(A_2)p(B/A_2) + p(A_n)p(B/A_n)}$$

14. Binomial Distribution

In an experiment, consisting of n independent trials in which an event has probability p in a single trial, the probability P_X of obtaining X successes is given by

$$P_X = C_{(n-X)}p^X q^{(n-X)}$$

where

$$q = (1 - p) \text{ and } C_{(n, X)} = \frac{n!}{X!(n-X)!}$$

The probability of P_x between a and b successes (both a and b included) is $P_a + P_{a+1} + \ldots + P_b$, so if $a = 0$ and $b = n$, the sum;

$$\sum_{X=0}^{n} C_{(n,X)} p^X q^{(n-X)} = q^n + C_{(n,1)} q^{n-1} p + C_{(n,2)} q^{n-2} p^2 + \ldots + p^n$$

$$= (q+p)^n = 1.$$

15. Mean of Binomially Distributed Variable

The mean number of successes in n independent trials is $m = np$ with standard deviation $\sigma = \sqrt{npq}$.

Normal Distribution

In the binomial distribution as n increases, the histogram of heights is approximated by the bell-shaped curve (normal curve)

$$Y = \frac{1}{\sigma \sqrt{2\pi}} e^{-(x-m)^2/2\sigma^2}$$

where, $m =$ the mean of the binomial distribution $= np$ and $\sigma\sqrt{npq}$ is the standard deviation. For any normally distributed random variable X with mean m and standard deviation σ, the probability function (density) is given by $\sqrt{npq}$.

The *standard* normal probability curve is given by

$$y = \frac{1}{\sqrt{2\pi}} e^{-Z^2/2}$$

and has mean $= 0$ and standard deviation $= 1$. The total area under the standard normal curve is 1. Any normal variable X can be put into standard form by defining $Z = (X - m)/\sigma$; thus the probability of X between the given X_1 and X_2 is the area under the standard normal curve between the corresponding Z_1 and Z_2. The standard normal curve is often used instead of the binomial distribution in experiments with discrete outcomes.

16. Poisson Distribution

$$P = \frac{e^{-m} m^r}{r!}$$

is an approximation to the binomial probability for r successes in n trials when $m = np$ is small (< 5) and the normal curve is not recommended to approximate binomial probabilities. The variance σ^2 in the Poisson distribution is np, the same value as the mean.

17. Probable Error (λ)

$$\text{Probable error } (\lambda) = 0.6745; \ \sigma = \frac{2}{3}\lambda \text{ (approx)}$$

18. Standard Error

Standard deviation of a sample means is called *standard error* (SE) $= \dfrac{\sigma}{\sqrt{n}}$.

19. Summary of Probability Distributions

Continuous Distributions

$$\textbf{Normal} \ \ y = \frac{1}{\sigma\sqrt{2\pi}} \exp\left[-(x-m)^2 / 2\sigma^2\right]$$

Mean $= m$, variance $= \sigma^2$

$$\textbf{Standard normal} \ \ y = \frac{1}{\sqrt{2x}} \exp\left[-z^2 / 2\right]$$

Mean $= 0$, variance $= 1$

F-Distribution

$$y = A \frac{F^{\frac{f_1-2}{2}}}{(f_2 + f_1 F)^{\frac{f_1+f_2}{2}}}$$

where,

$$A = \frac{\Gamma\left(\dfrac{f_1+f_2}{2}\right)}{\Gamma\left(\dfrac{f_1}{2}\right)\Gamma\left(\dfrac{f_2}{2}\right)} f_1^{\frac{f_1}{2}} f_2^{\frac{f_2}{2}}$$

$$\text{Mean} = \frac{f_2}{f_2 - 2}, \text{ variance} = \frac{2 f_2^2 (f_1 + f_2 - 2)}{f_1 (f_2 - 2)^2 (f_2 - 4)}$$

Chi-square Test

$$y = \frac{1}{2^{f/2} \Gamma(f/2)} \exp\left(-\frac{1}{2} x^2\right) (x^2)^{\frac{f-2}{2}}$$

Mean = f, variance = $2f$

Students t-Test

$$y = A(1 + t^2/f)^{-(f+1)/2}; \text{ where } A = \frac{\Gamma(f/2 + 1/2)}{\sqrt{f} \, \pi \, \Gamma(f/2)}$$

$$\text{Mean} = 0, \text{ Variance} = \frac{f}{f - 2} \ (\text{for } f > 2)$$

Discrete Distribution

Binomial distribution $y = C_{(n, x)} p^x (1 - p)^{n - x}$

Mean = np, variance = $np \, (1 - p)$

Poisson Distribution

$$y = \frac{e^{-m} m^x}{x!}$$

Mean = m, variance = m^x

Chapter **10**

Coordinate Geometry

1. Rectangular Coordinates

The point in a plane may be placed in one-to-one correspondence with pairs of real numbers. A common method is to use perpendicular lines that are horizontal and vertical and intersect at a point called the *origin*. These two lines constitute the coordinate axes; the horizontal line is the x-axis and the vertical line is the y-axis. The positive direction of the x-axis is to the right whereas the positive direction of the y-axis is up. If P is a point in the plane, one may draw lines through it that are perpendicular to the x- and y-axes (such as the broken lines of Fig. 10.1). The lines intersect the x-axis at a point with coordinate

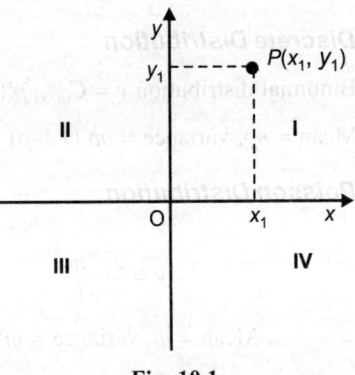

Fig. 10.1

x_1 and the y-axis at a point with coordinate y_1. We call x_1 the x-coordinate or *abscissa* and y_1 is termed by the y-coordinate or *ordinate* of the point P. Thus, point P is associated with the pair of real numbers (x_1, y_1) and is denoted by $P(x_1, y_1)$. The coordinate axes divide the plane into quadrants I, II, III and IV.

2. Distance between Two Points (Slope)

The distance d between the two points $P_1(x_1, y_1)$ and $P_2(x_2, y_2)$,
$$d = \sqrt{(x_2 - x_1)^2 + (y_2 - y_1)^2}$$

The midpoint of the line segment P_1P_2 is

$$\left(\frac{x_1 + x_2}{2}, \frac{y_1 + y_2}{2}\right).$$

The slope of the line segment P_1P_2, provided it is not vertical is denoted by m and is given by

$$\frac{y_2 - y_1}{x_2 - x_1}.$$

The slope is related to the angle of inclination α as shown in Fig. 10.2.

$$m = \tan \alpha$$

Two lines or line segments with slopes m_1 and m_2 are perpendicular if

$$m_1 \cdot m_2 = -1 \text{ or } m_1 = -1/m_2$$

and are parallel if $m_1 = m_2$

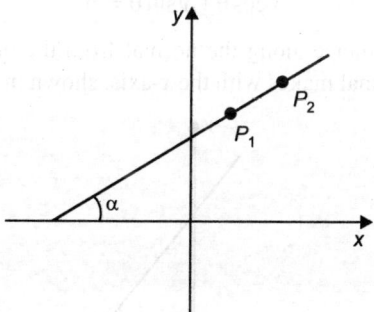

Fig. 10.2

3. Equations of Straight Lines

Point Slope Form: A *vertical* line has an equation of the form

$$x = c$$

where $(c, 0)$ is its intersection with the x-axis. A line of slope m through point (x_1, y_1) is given by

$$y - y_1 = m(x - x_1)$$

Thus, a *horizontal line* (slope $= 0$) through point (x_1, y_1) is given by $y = y_1$.

Two Point Form: A nonvertical line through the two points $P_1(x_1, y_1)$ and $P_2(x_2, y_2)$ is given by either

$$y - y_1 = \left(\frac{y_2 - y_1}{x_2 - x_1}\right)(x - x_1)$$

or

$$y - y_2 = \left(\frac{y_2 - y_1}{x_2 - x_1}\right)(x - x_2).$$

Intercept Form: A line with x-intercept a and y-intercept b is given by

$$\frac{x}{a} + \frac{y}{b} = 1 \qquad (a \neq 0, b \neq 0).$$

General Equation: The *general equation* of a line is

$$Ax + By + C = 0$$

Normal Form: The *normal form* of a straight line equation

$$= x\cos\theta + y\sin\theta = p$$

where p is the distance along the normal from the origin and θ is the angle that the normal makes with the x-axis, shown in Fig. 10.3.

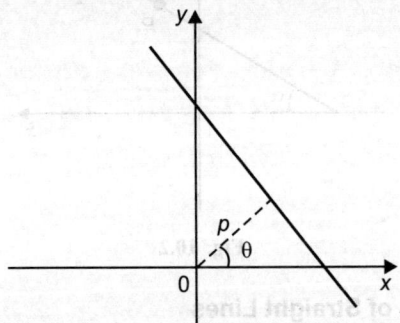

Fig. 10.3

The general equation of the line $Ax + By + C = 0$ may be written in normal form by dividing by $\pm\sqrt{A^2 + B^2}$, where the plus sign is used when C is negative and the minus sign is used when C is positive:

$$\frac{Ax + By + C}{\pm\sqrt{A^2 + B^2}} = 0.$$

so that

$$\cos\theta = \frac{A}{\pm\sqrt{A^2+B^2}} \qquad \sin\theta = \frac{B}{\pm\sqrt{A^2+B^2}}$$

and

$$p = \frac{|C|}{\sqrt{A^2+B^2}}$$

4. Distance from a Point to a Line

The perpendicular distance from a point $P(x_1, y_1)$ to the line $Ax + By + C = 0$ is given by

$$d = \frac{Ax_1 + By_1 + C}{\pm\sqrt{A^2+B^2}}.$$

5. Circle

The equation of a circle of radius r and centre at $P(x_1, y_1)$ is

$$(x - x_1)^2 + (y - y_1)^2 = r^2$$

General equation of circle is $x^2 + y^2 + 2gx + 2fy + c = 0$

Its centre is $(-g, -f)$ and radius $\sqrt{g^2 + f^2 - c}$, where g and f are half the coefficients of x and y and c is a constant term.

6. Parabola

A parabola is the set of all points (x, y) in the plane that are equidistant from a given line called the *directrix* and a given point called the *focus*. The parabola is symmetric about a line that contains the focus and is perpendicular to the directrix. The line of symmetry intersects the parabola at its *vertex* (Fig. 10.4). The eccentricity $e = 1$.

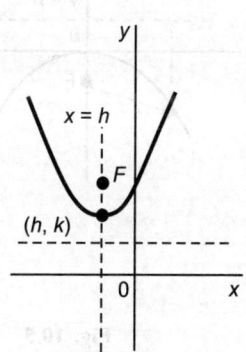

Fig. 10.4

The distance between the focus and the vertex, or vertex and directrix, is denoted by $p\ (> 0)$ and leads to one of the following equations of parabola with vertex at the origin (Figs 10.5 and 10.6).

$$y = \frac{x^2}{4p} \quad \text{(opens upward)}$$

$$y = -\frac{x^2}{4p} \text{ (opens downward)}$$

$$x = \frac{y^2}{4p} \text{ (opens to right)}$$

$$x = -\frac{y^2}{4p} \text{ (opens to left)}$$

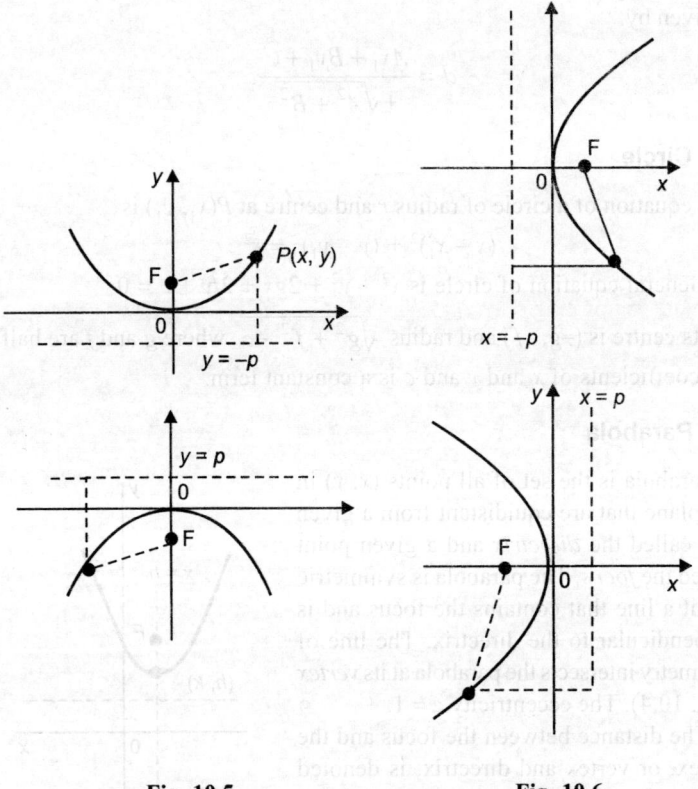

Fig. 10.5   **Fig. 10.6**

7. Orthogonal Spheres

Two spheres are said to be orthogonal if the tangent planes at a point of intersection are at right angles.

8. Ellipse

An ellipse is the set of all points in the plane such that the sum of their distances from two fixed points, called *foci*, is a given constant $2a$. The distance between the foci is donated $2c$; the length of the major axis is $2a$, whereas the length of the minor axis is $2b$ (Fig. 10.7).

$$a = \sqrt{b^2 + c^2}.$$

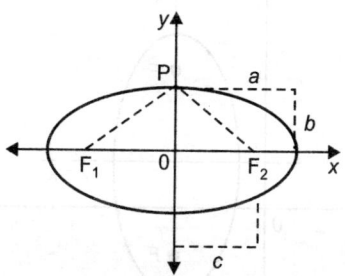

Fig. 10.7

The eccentricity of an ellipse e is <1. An ellipse with centre at point (h, k) and major axis *parallel to the x-axis* (Fig. 10.8) is given by the equation

$$\frac{(x-h)^2}{a^2} + \frac{(y-k)^2}{b^2} = 1$$

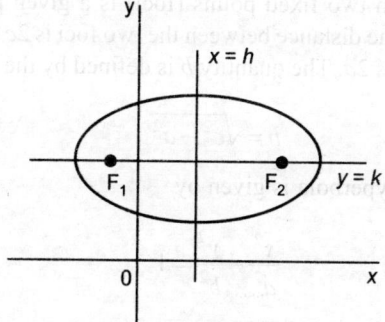

Fig. 10.8

$$\frac{(y-h)^2}{a^2} + \frac{(y-k)^2}{b^2} = 1$$

An ellipse with centre at (h, k) and major axis parallel to the y-axis is given by the equation (Fig. 10.9).

$$\frac{(y-k)^2}{a^2} + \frac{(x-h)^2}{b^2} = 1$$

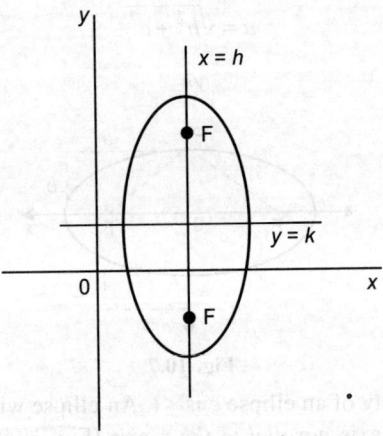

Fig. 10.9

9. Hyperbola (e > 1)

A hyperbola is the set of all points in the plane such that the difference of its distances from two fixed points (foci) is a given positive constant denoted by $2a$. The distance between the two foci is $2c$ and that between the two vertices is $2a$. The quantity b is defined by the equation

$$b = \sqrt{c^2 - a^2}$$

Equation of hyperbola is given by

$$\frac{x^2}{a^2} - \frac{y^2}{b^2} = 1$$

10. Change of Axes

A change in the position of the coordinate axes will generally change the coordinates of the points in the plane. The equation of a particular curve will also generally change.

11. Translation

When the new axes remain parallel to the original, the transformation is called a *translation* (Fig. 10.10). The new axes, denoted by x' and y', have $0'$ at (h, k) with reference to the x and y axes.

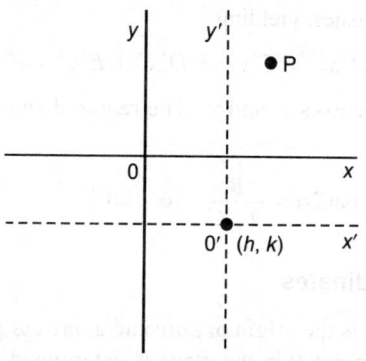

Fig. 10.10

A point P with coordinates (x, y) with respect to the original has coordinates (x', y') with respect to the new axes. These are related by

$$x = x' + h$$

$$y = y' + k$$

12. Rotation

When the new axes are drawn through the same origin, remaining mutually perpendicular, but titled with respect to the original, the transformation is one of rotation. For angle of rotation ϕ (Fig. 10.11), the coordinates (x, y) and (x', y') of point P are related by

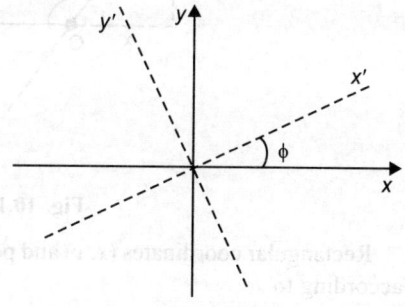

Fig. 10.11

$$x = x' \cos \phi - y' \sin \phi$$

$$y = x' \sin \phi + y' \cos \phi$$

13. General Equation of Degree Two

$$Ax^2 + Bxy + Cy^2 + Dx + Ey + F = 0$$

Every equation of the above form defines a conic section or one of the limiting forms of a conic. By rotating the axes through a particular angle ϕ, the xy-term vanishes, yielding

$$A'x'^2 + C'y'^2 + D'x' + E'y' + F' = 0$$

with respect to the axes x' and y'. The required angle ϕ (Fig. 10.11) is calculated from

$$\tan 2\phi = \frac{B}{A - C} \quad (\phi < 90°).$$

14. Polar Coordinates

The fixed point O is the origin or *pole* and a line OA drawn through it is the polar axis. A point P in the plane is determined from its distance r measured from O, and the angle θ from the pole are positive, whereas those measured in the opposite direction are negative as shown in Fig. 10.12.

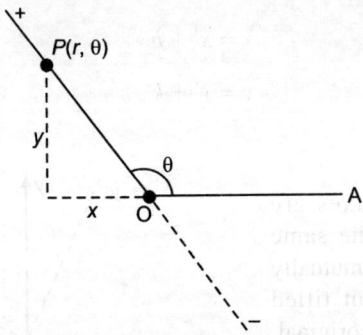

Fig. 10.12

Rectangular coordinates (x, y) and polar coordinates (r, θ) are related according to

$$x = r \cos \theta, \qquad y = r \sin \theta$$

$$r^2 = x^2 + y^2, \qquad \tan \theta = y/x$$

Several well-known polar curves are shown in Fig. 10.13 (a, b, c, d, e).

The polar equation of a comic section with focus at the pole and distance $2p$ from directrix to focus is either

$$r = \frac{2ep}{1 - e\cos\theta} \qquad \text{(directrix to left of pole)}$$

or

$$r = \frac{2ep}{1 + e\cos\theta} \qquad \text{(directrix to right of pole)}$$

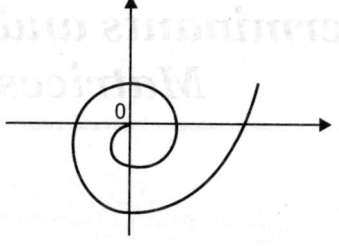

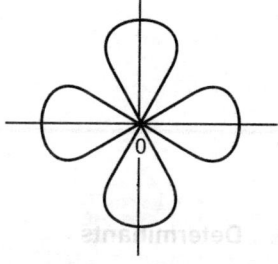

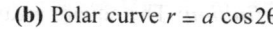

(a) Polar curve $r = e^{a\theta}$ **(b)** Polar curve $r = a\cos 2\theta$

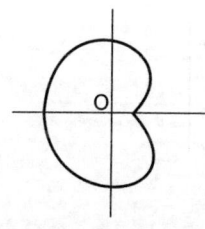

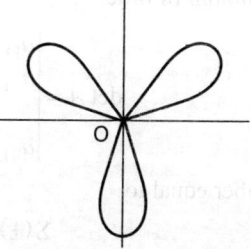

(c) Polar curve $r = 2a\cos\theta + b$ **(d)** Polar curve $r = a\sin 3\theta$

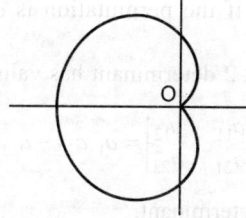

(e) Polar curve $r = a(1 - \cos\theta)$

Fig. 10.13 Various types of polar curves

The corresponding equations for the directrix below or above the pole are same, except that $\sin\theta$ appears instead of $\cos\theta$.

Chapter 11

Determinants and Matrices

---

1. Determinants

Definition: The square array A, with n rows and n columns is called a *determinant* of order n.

$$\det A = \begin{vmatrix} a_{11} & a_{12} & ... & a_{1n} \\ a_{21} & a_{22} & ... & a_{2n} \\ ... & ... & ... & ... \\ a_{n1} & a_{n2} & ... & a_{nn} \end{vmatrix},$$

a number equal to

$$\Sigma(\pm)a_{1i}\, a_{2j}\, a_{3k} ... a_{nl}$$

where $i, j, k, ..., l$ is the permutation of the n integers $1, 2, 3,..., n$ in some order. The sign is plus if the permutation is *even* and minus if the permutation is *odd*.

For example: The 2×2 determinant has value

$$\Delta = \begin{vmatrix} a_{11} & a_{12} \\ a_{21} & a_{22} \end{vmatrix} = a_{11}a_{22} - a_{12}a_{21}$$

Δ denotes the value of determinant.

Note:

    a. The number $a_{11}, a_{12}, a_{13}, a_{21}, a_{22}, a_{23}, a_{31}, a_{32}, a_{33}$, etc. are called *elements of the determinants.*

    b. Horizontal lines are called *rows* and vertical lines as *columns*.

   c. The elements in the diagonal line namely a_{11}, a_{22}, a_{33} are called *principal diagonal.*

   d. Determinants of the nth order have n rows and n columns.

2. Calculation of Cofactors

Each element a_{ij} has a determinant of order $(n-1)$ called a *minor* (M_{ij}) obtained by suppressing all elements in row i and column j. For example, the minor of element a_{22} in the 3×3 determinant is

$$\begin{vmatrix} a_{11} & a_{13} \\ a_{31} & a_{33} \end{vmatrix}$$

The cofactor of element a_{ij} denoted A_{ij} is defined as $\pm M_{ij}$, where the sign is determined from i and j:

$$A_{ij} = (-1)^{i+j} \, M_{ij}$$

The value of the $n \times n$ determinant equals to the sum of products of elements of any row (or column) and their respective cofactors. Thus, the 3×3 determinant A is given by

$$\det A = \begin{vmatrix} a_{11} & a_{12} & a_{13} \\ a_{21} & a_{22} & a_{23} \\ a_{31} & a_{32} & a_{33} \end{vmatrix} \text{ is called a } \textit{determinant of the third order.} \text{ Its}$$

value Δ

$$= a_{11} \begin{vmatrix} a_{22} & a_{23} \\ a_{32} & a_{33} \end{vmatrix} - a_{12} \begin{vmatrix} a_{21} & a_{23} \\ a_{31} & a_{33} \end{vmatrix} + a_{13} \begin{vmatrix} a_{21} & a_{22} \\ a_{31} & a_{32} \end{vmatrix}$$

$$= a_{11}(a_{22}\,a_{33} - a_{23}\,a_{32}) - a_{12}(a_{21}\,a_{33} - a_{23}\,a_{31}) + a_{13}(a_{21}\,a_{32} - a_{31}\,a_{22})$$

$$\det A = a_{11}A_{11} + a_{12}A_{12} + a_{13}A_{13} \text{ (first row)}$$

or

$$= a_{11}A_{11} + a_{21}A_{21} + a_{31}A_{31} \text{ (first column) etc.}$$

3. Properties of Determinants

   a. If the corresponding columns and rows of A are interchanged, value of determinant A is unchanged.

   b. If two adjacent rows or columns are interchanged, the sign of determinant A changes.

   c. If any two rows or columns are identical, then determinant $A = 0$.

d. If A is triangular (all elements above the main diagonal equal to zero), $A = a_{11} \cdot a_{22} \ldots a_{mn}$:

$$= \begin{vmatrix} a_{11} & 0 & 0 & \ldots & 0 \\ a_{21} & a_{22} & 0 & \ldots & 0 \\ \ldots & \ldots & \ldots & \ldots & \ldots \\ a_{n1} & a_{n2} & a_{n3} & \ldots & a_{mn} \end{vmatrix}$$

e. If to each element of a row or column there is added K times the corresponding element in another row or column, the value of the determinant is unchanged.

f. If each element in any row or column is multiplied by the same number, the whole determinant is multiplied by that number.

g. The product of two determinant of the same order is a determinant of the same order.

4. Matrices

Definition: A matrix is a rectangular array of numbers and represented by a symbol A or $[a_{ij}]$:

$$A = \begin{bmatrix} a_{11} & a_{12} & \ldots & a_{1n} \\ a_{21} & a_{22} & \ldots & a_{2n} \\ \ldots & \ldots & \ldots & \ldots \\ a_{m1} & a_{m2} & \ldots & a_{mn} \end{bmatrix} = [a_{ij}]$$

The numbers a_{ij} are termed *elements* of the matrix; subscripts i and j identify the element as the number in row i and column j. The order of the matrix is $m \times n$ ("m by n"). When $m = n$, the matrix is square and is said to be of order n. For a square matrix of order n, the elements a_{11}, a_{22},...., a_{mn} constitute the main diagonal.

5. Operations

Addition: Matrices A and B of the same order may be added by adding corresponding element, *i.e.* $A + B = [(a_{ij} + b_{ij})]$.

Scalar multiplication: If $A = [a_{ij}]$ and K is a constant (scalar), then $KA = [Ka_{ij}]$, that is, every element of A is multiplied by K.

Multiplication of matrices: Matrices A and B may be multiplied when, the number of columns of A equals the number rows of B.

For example, if

$$
\begin{bmatrix}
a_{11} & a_{12} & \cdots & a_{1k} \\
a_{21} & a_{22} & \cdots & a_{2k} \\
\cdots & \cdots & \cdots & \cdots \\
a_{m1} & \cdots & \cdots & a_{mk}
\end{bmatrix}
\begin{bmatrix}
b_{11} & b_{12} & \cdots & b_{1n} \\
b_{21} & b_{22} & \cdots & b_{2n} \\
\cdots & \cdots & \cdots & \cdots \\
b_{k1} & b_{k2} & \cdots & b_{kn}
\end{bmatrix}
$$

$$
=
\begin{bmatrix}
c_{11} & c_{12} & \cdots & c_{1n} \\
a_{21} & a_{22} & \cdots & a_{2n} \\
\cdots & \cdots & \cdots & \cdots \\
a_{m1} & c_{m2} & \cdots & a_{mn}
\end{bmatrix}
$$

then element c_{21} is the sum of products $a_{21}\, b_{11} + a_{22}\, b_{21} + \ldots + a_k\, b_{kl}$.

6. Transpose

If A is an $n \times m$ matrix, the matrix of order $m \times n$ obtained by interchanging the rows and columns of A is called the *transpose* and is denoted A^T. The following are properties of A, B, and their respective transposes:

$$(A^T)^T = A$$

$$(A + B)^T = A^T + B^T$$

$$(KA)^T = KA^T$$

$$(AB)^T = B^T A^T$$

A *symmetric* matrix is a square matrix A with the property $A = A^T$

7. Unit Matrix

$$
\begin{bmatrix}
1 & 0 & 0 & \cdots & 0 \\
0 & 1 & 0 & \cdots & 0 \\
0 & 0 & 1 & \cdots & 0 \\
\cdots & \cdots & \cdots & \cdots & \cdots \\
0 & 0 & 0 & \cdots & 1
\end{bmatrix}
$$

A scalar matrix with diagonal elements 1 is called the *identity* or *unit matrix* and is denoted I. Thus, for any nth order matrix A, $AI = IA = A$.

8. Null Matrix

A matrix rectangular or square, each of whose elements are zero is called a *zero* or *null matrix*.

9. Triangular Matrices

A square matrix $A = (a_{ij})_{n \times n}$ is called *upper triangular matrix* if $a_{ij} = 0$ for $i > j$ and is called *lower triangular matrix* if $a_{ij} = 0$ for $i < j$.

10. Submatrix

A matrix obtained by deleting some rows or columns or both of a given matrix is called *submatrix* of a given matrix.

11. Symmetric Matrices

A symmetric matrix is a special kind of square matrix $A = [a_{ij}]$ for which $a_{ij} = a_{ji}$ for all i and j.

12. Skew Symmetric Matrix

It is a square matrix A if $A^T = -A$.

13. Complex Conjugate of a Matrix

It is a matrix obtained by replacing all its elements by their respective conjugates.

14. Adjoint Matrix

If A is an n-order square matrix and A_{ij} the cofactor of element a_{ij}, the transpose of $[A_{ij}]$ is called the *adjoint* of matrix A.

$$\text{adj } A = [A_{ij}]^T$$

15. Inverse Matrix

Given a square matrix A of order n, if there exists a matrix B such that $AB = BA = I$, then B is called the *inverse of A*. The inverse is denoted A^{-1}. A necessary and sufficient condition that the square matrix A have an inverse is det $A \neq 0$. Such a matrix is called *nonsingular*; its inverse is unique and it is given by

$$A^{-1} = \frac{\text{adj } A}{\text{det } A}$$

16. Matrix Solution

The linear system may be written in matrix form $AX = B$ where, A is the matrix of coefficients $[a_{ij}]$, X and B are given by

$$X = \begin{bmatrix} x_1 \\ x_2 \\ \cdot \\ \cdot \\ \cdot \\ x_n \end{bmatrix} \qquad B = \begin{bmatrix} b_1 \\ b_2 \\ \cdot \\ \cdot \\ \cdot \\ b_n \end{bmatrix}$$

If a unique solution exists, det $A \neq 0$; hence A^{-1} exists and $X = A^{-1} B$.

Chapter 12

Electromagnetics

1. Unit Vectors and Coordinate Systems

The unit vectors for the cartesian (rectangular) system shown in Fig. 12.1(a) are: a_x, a_y, a_z and all three vectors are constant.

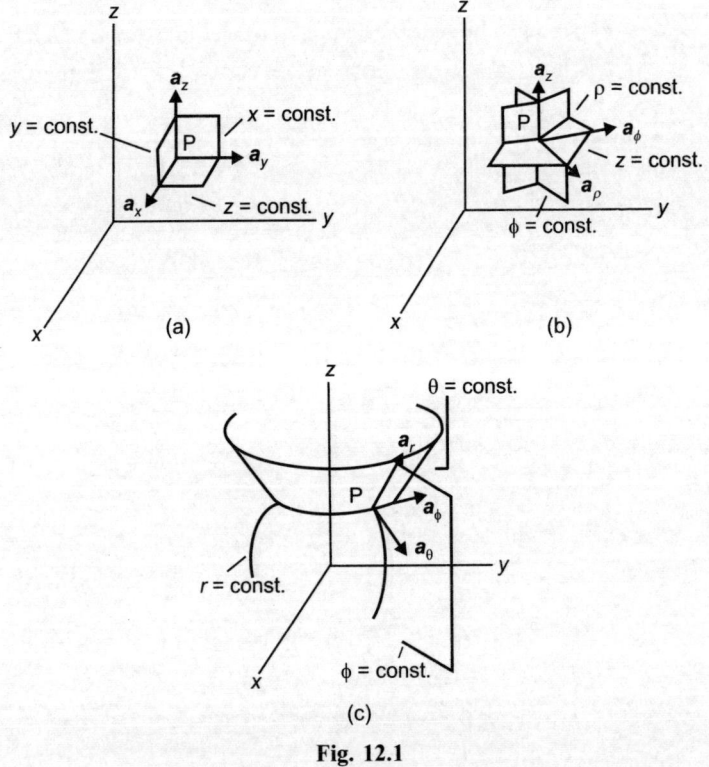

Fig. 12.1

The unit vectors for the Cylindrical coordinate system shown in Fig. 12.1b are: a_p, a_ϕ, a_z where, a_p, a_ϕ and a_z are constant.

The unit vectors for the spherical coordinate system shown in Fig. 12.1c are: a_r, a_θ, a_ϕ.

2. Electric Field Intensity

The electrostatic field intensity is defined as the force on Q when $Q = 1$ C. Thus

$$E = \frac{Q_1}{4\pi\varepsilon_0 R^2} a_R \text{ V/m}$$

and
$$F = QE \text{ newton}$$

3. Coulomb's Laws

The force (F) between two charges q_1 and q_2 is

  (i) directly proportional to the product of the charges q_1 and q_2;

 (ii) inversely proportional to the square of distance d between them;

(iii) depends on the nature of medium surrounding the charges.

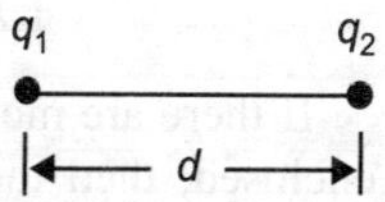

Mathematically,

$$F \propto \frac{q_1 \cdot q_2}{d^2}$$

$$F = \frac{q_1 \cdot q_2}{4\pi\varepsilon_r \varepsilon_0 d^2} a_R \text{ newton}$$

Fig. 12.2

where ε_0 is the permittivity of air and its value is 8.854×10^{-12} F/m and ε_r is the relative permittivity of surrounding medium with respect to air and a_R the unit vector in the direction of line joining the two charges.

4. Electric Field of Many Charges

If there are several point charges q_1, q_2, q_3,..., q_n located at different points. The electric field intensity at point P is $E = \dfrac{F}{q}$

$$= \frac{q_1}{4\pi\varepsilon_0 R_1^2} i_{R_1} + \frac{q_2}{4\pi\varepsilon_0 R_2^2} i_{R_2} + ... + \frac{q_n}{4\pi\varepsilon_0 R_n^2} i_{R_n}$$

$$= \sum_{j=1}^{n} \frac{q_j}{4\pi\varepsilon_0 R_j^2} i_{R_j}$$

5. Gauss Law of Electricity

The surface integral of the normal component of electric field intensity E over a *closed surface containing point charge q*, is given by

$$\oint E \cdot dS = q/\varepsilon_0$$

This can be interpreted as the net flux of electric field emanating from the surfaces S containing a point charge q is equal to q/ε_0. If this arbitrary surface does not enclose the point charge, the net electric field flux emanating from the surface must be zero, *i.e.*

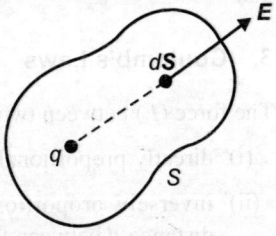

$$\oint E \cdot dS = 0$$

Fig. 12.3

If there are more than one point charges enclosed, then the above equation can be generated as follows:

$$\oint_s E \cdot dS = \oint_s E_1 \cdot dS + \oint_s E_2 \cdot dS + \oint_s E_3 \cdot dS + ...$$

$$= \frac{q_1 + q_2 + ... + q_n}{\varepsilon_0} = \frac{\text{charge enclosed by the surface } S}{\varepsilon_0}$$

6. Gauss Law in Differential Form

Consider a volume distribution with the charge density ρ. The charge enclosed by arbitrary closed surface S is given by volume integral of charge density throughout the volume V enclosed by surface, that is, $\int_V \rho dv$. According to Gauss law

$$\oint_s E \cdot dS = \frac{1}{\varepsilon_0} \int_V \rho dv$$

If the volume is shrunk to a very small Δv, the surface area becomes very small ΔS.

$$\underset{\Delta V \to 0}{\text{Lim}} \oint_{\Delta S} \frac{E \cdot dS}{\Delta v} = \underset{\Delta V \to 0}{\text{Lim}} \frac{\left(\frac{1}{\varepsilon_0}\right) \oint \rho \, dv}{\Delta v}$$

$$= \frac{1}{\varepsilon_0} \underset{\Delta v \to 0}{\text{Lim}} \frac{\rho \Delta v}{\Delta v} = \frac{1}{\varepsilon_0} \cdot \rho$$

or
$$\nabla E = \frac{1}{\varepsilon_0} \rho$$

This equation is *Gauss law* in differential form. It states that the divergence of electric field intensity at any point is equal to $\dfrac{1}{\varepsilon_0}$ times the volume charge density at that point. This is *Maxwell's divergence equation* for electric field.

7. Electric Potential

It is defined as the work done W_{AB} by the field in moving a rest charge q from A to B along a given path

$$W_{AB} = q \int_A^B E \cdot dl$$

8. Potential due to Group of Charges

The potential at a point due to a group of point charges $q_1, q_2, ..., q_n$ is the algebraic sum of the potentials due to each charge. That is

$$V = \frac{1}{4\pi\varepsilon_0} \left(\frac{q_1}{r_1} + \frac{q_2}{r_2} + ... + \frac{q_n}{r_n} \right)$$

$$= \frac{1}{4\pi\varepsilon_0} \cdot \sum_n \frac{q_n}{r_n}$$

9. Electric Potential Energy

The workdone required to move q_2 from infinity to distance r by definition of potential will be

$$W = V q_2$$

Therefore, electric potential energy $U = V q_2$

$$= \frac{1}{4\pi\varepsilon_0} \cdot \frac{q_1 q_2}{r}$$

10. Poisson's Equation

It expresses the relationship of potential at a point to the volume charge density ρ at that point,

$$\nabla^2 V = \frac{\rho}{\varepsilon_0}$$

If the volume charge density is zero in a region, then $\nabla^2 V = 0$
This is known as *Laplace's equation.*

11. Current Density

The current density J is related to the electric field E for a metallic conductor as

$$J = \sigma E$$

where σ is the conductivity of the conductor.

The current density J is a convection current,

$$J = \rho v$$

where v is a velocity vector and ρ is the volume charge density.

12. Biot-Savart's Law

A current I flowing in a differential vector length dL results in a magnetic field intensity H as

$$dH = \frac{IdL \times a_R}{4\pi R^2} \text{ A/m}$$

Expressed in terms of current density J, we have

$$H = \int_{\text{volume}} \frac{J \times a_R \, dv}{4\pi R^2}$$

13. Maxwell's Equations for Static Fields

$$\nabla \times H = J$$

and $\qquad\qquad \nabla \times E = 0$

14. Stoke's Theorem

$$\oint H \cdot dL = \int_{\text{surface}} (\nabla \times H) \cdot dS$$

15. Magnetic Flux Density

Magnetic flux density B in free space is

$$B = \mu_0 \, H \; \text{T}$$

where T is tesla and $\mu_0 = 4\pi \times 10^{-7}$ H/m.

Then, the divergence theorem provides

$$\nabla \cdot B = 0$$

16. Maxwell's Equations for Statistics Fields

Differential Form	Integral Form
$\nabla \cdot D = \rho$	$\oint_{\text{surface}} D \cdot dS = Q = \int_{\text{volume}} \rho \, dv$
$\nabla \times E = 0$	$\oint E \cdot dL = 0$
$\nabla \times H = J$	$\oint H \cdot dL = I = \int_{\text{surface}} J \cdot dS$
$\nabla \cdot B = 0$	$\oint B \cdot dS = 0$

17. Maxwell's Equations for Time-varying Fields

$$\nabla \times E = -\frac{\partial B}{\partial t}$$

$$\nabla \times H = J + \frac{\partial D}{\partial t}$$

$$\nabla \cdot D = \rho$$

$$\nabla \cdot B = 0$$

18. Magnetic Deflecting Force

If the positive test charge q_0 is fired with velocity v through a point P in a magnetic field B, then the force F acting on the moving charge is given by

$$F = q_0 v \times B$$

i.e.

$$F = q_0 \cdot v \, B \sin \theta$$

where θ is the angle between v and B. The force experienced by the moving charge due to the magnetic field is normal to both v and B. Therefore, there is no acceleration along the direction of motion and the magnetic field can only change the direction (Fig. 12.4).

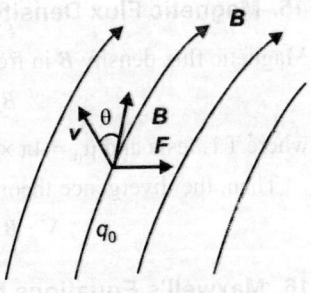

Fig. 12.4

19. Lorentz Relation

When a charged particle q_0 is subjected to an electric field E and magnetic field B, the resultant force acting on it is given by

$$F = q_0 E + q_0 v \times B$$

For a continuous charge distribution of density ρ moving with a velocity v, the force per unit volume can be defined as

$$F = \rho E + J \times B$$

where J is the volume current density and is given by $J = \rho v$

20. Magnetic Flux Density

F_1 and F_2 are two non-zero forces for two velocities v_1 and v_2 in different directions in the magnetic field B.

$$F_1 \text{ and } F_2 = (q v_1 \times B) \times (q v_2 \times B)$$

or

$$B = \frac{F_2 \times F_1}{q (F_1 \cdot v_2)}$$

21. Force on a Conductor

Force on a current carrying conductor in magnetic field is given by

$$F = lI \times B$$

$$= BIl \sin \theta$$

The direction of force can be determined by applying Flemming's left hand rule. The thumb, fore finger and middle finger of the left hand are stretched in such a way that they are at right angles to each other mutually and the fore finger points towards the direction of the magnetic field, middle finger towards the direction of the flow of current, then thumb will be in the direction of force acting on the conductor (Fig. 12.5).

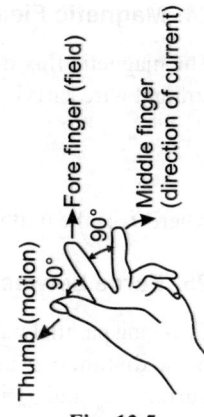

Fig. 12.5

22. Force on a Current Carrying Conductor

The total force experienced by a filamentary wire carrying current I as shown in Fig. 12.6 is given by

$$F = \int_{\text{wire}} (I d\mathbf{l} \times \mathbf{B})$$

$$= I \int_{\text{wire}} (d\mathbf{l} \times \mathbf{B})$$

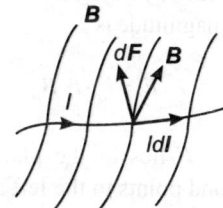

Fig. 12.6

23. Biot-Savart Law (Integral Form)

The field strength due to a small conductor of length δl carrying current I at any point is

(i) directly proportional to length δl;

(ii) directly proportional to the strength of current flowing;

(iii) inversely proportional to the square of the perpendicular distance of the point from δl.

Mathematically, $dB = \dfrac{\mu_0 I \delta l}{4\pi R^2} \cdot i_R$

The magnetic flux density due to a wire of any length is given by

$$B = \frac{\mu_0}{4\pi} \int_C \frac{I d\mathbf{l} \times i_R}{R^2}$$

where the integral is taken along the contour C of the wire.

24. Magnetic Field due to a Current Carrying Wire

The magnetic flux density at point P at a distance r from infinitely long straight wire carrying current I is:

$$B = \frac{\mu_0 I}{2\pi r} i_\phi$$

where i_ϕ is the unit vector in the direction of magnetic field.

25. Force between two Parallel Conductors

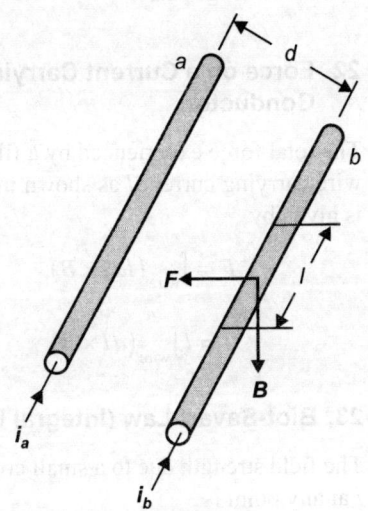

Two long parallel wires separated by a distance d and carrying currents i_a and i_b in a length l of the wire, it will experience a sideways magnetic force whose magnitude is

$$F = i_a \cdot l \cdot B_a = \frac{\mu_0 l\, i_a\, i_b}{2\pi d}$$

F lies in the plane of wires and points to the left as shown in Fig. 12.7.

The forces that the two wires exert on each other are equal and opposite. For anti-parallel currents, the two wires will repel each other.

Fig. 12.7

26. The Ampere's Law

Consider a circular path C of radius r in the plane normal to the wire and centred at the wire as shown in Fig. 12.8. Take an infinitesimal element of length dl on this path. Then

$$\oint_C B \cdot dl = \int_C \frac{\mu_0 I\, dl}{2\pi r} = \frac{\mu_0 I}{2\pi r} \cdot 2\pi r$$

$$= \mu_0 I$$

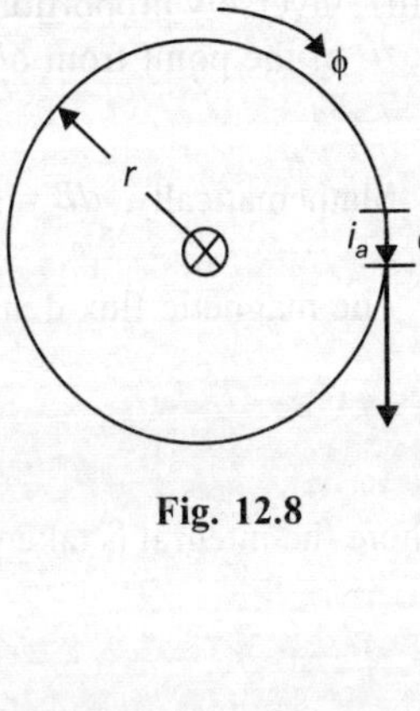

Fig. 12.8

When generalized over the arbitrary closed path around the wire, this equation is known as Ampere's law. It gives information about the current enclosed by the path C. If the arbitrary path does not enclose current, then

$$\oint_C B \cdot dl = 0$$

27. Ampere's Law in Differential Form

The current enclosed by arbitrary closed path C is given by the surface integral of the current density over any surface S bounded by closed path C. That is

$$I = \int_S J \cdot dS$$

28. Capacitance of Various Systems

Capacitance of an isolated sphere of radius r in a medium of relative permittivity ε_r

$$= 4\pi\varepsilon_0\varepsilon_r \cdot r \text{ farad}$$

Capacitance of a spherical shell with outer sphere earthed as shown in Fig. 12.9a.

$$= 4\pi\varepsilon_0\varepsilon_r \frac{ab}{b-a} \text{ farad}$$

Spherical capacitor with inner sphere earthed as shown in Fig. 12.9b

$$= 4\pi \varepsilon_0 \varepsilon_r \, ab/(b-a) + 4\pi \varepsilon_0 \, b \text{ farad}$$

Capacitor of uniform parallel plates with medium partly air as shown in Fig. 12.9c.

$$= \left[\frac{\varepsilon_0 A}{d - \left(t - \dfrac{t}{\varepsilon_r}\right)} \right] \text{farad}$$

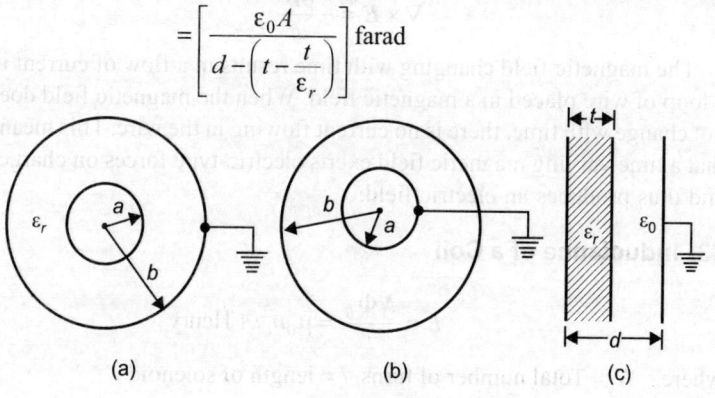

(a)   (b)   (c)

Fig. 12.9

29. Energy Stored in a Capacitor

$$W(=U) = \frac{1}{2}CV^2$$

30. Faraday's Laws of Electromagnetic Induction

First Law: This law states that when the flux linking with the coil or circuit changes, an *emf* is induced in it.

Second Law: This law states that the magnitude of *emf* induced is directly proportional to the rate of change of flux linkage. That is

$$\text{Induced emf} \propto N\frac{d\Phi}{dt}$$

31. Lenz's Law

This law states that the direction of induced *emf* is such that the current produced by it set up a magnetic field opposing the motion.

32. Faraday's Law in Integral Form

The magnetic flux enclosed by C is given by the surface integral of the magnetic flux density evaluated over the surfaces bounded by the contour C. Thus, Faraday's law can be defined as

$$\oint_c \boldsymbol{E} \cdot d\boldsymbol{l} = -\frac{d}{dt}\int_s \boldsymbol{B} \cdot d\boldsymbol{S}$$

Faraday's law in differential form is given by

$$\nabla \times \boldsymbol{E} = -\frac{\partial \boldsymbol{B}}{dt}$$

The magnetic field changing with time results in a flow of current in a loop of wire placed in a magnetic field. When the magnetic field does not change with time, there is no current flowing in the wire. This means that a time varying magnetic field exerts electric-type forces on charges and thus produces an electric field.

33. Inductance of a Coil

$$L = \frac{N\Phi_B}{i} = \mu_0 n^2 lA \text{ Henry}$$

where, N = Total number of turns, l = length of solenoid
             n = number of turns per unit length, A = cross-sectional area

34. Inductance in Series

When the coils are connected in series as shown in Fig. 12.10 such that their fluxes are additive, then the equivalent inductance is given by $L = L_1 + L_2 + 2M$, where M is the coefficient of mutual inductance. When the fluxes oppose each other, the equivalent inductance will be $L = L_1 + L_2 - 2M$.

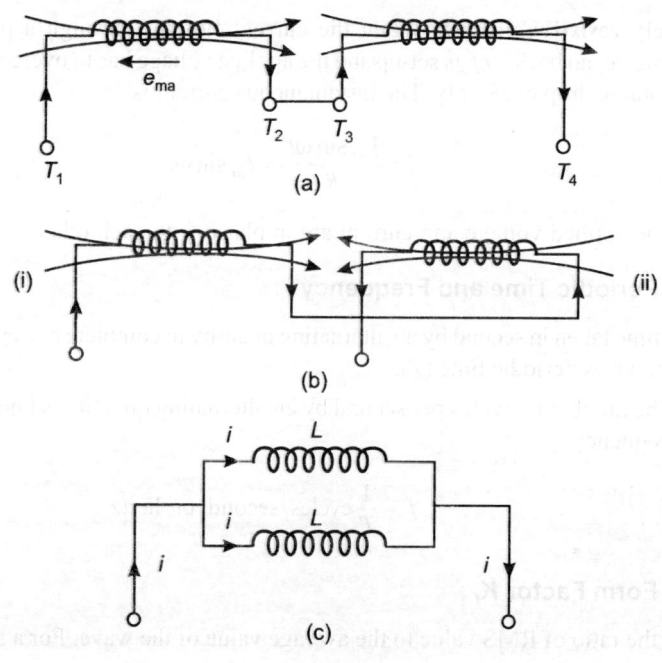

Fig. 12.10

35. Inductance in Parallel

The equivalent inductance, when mutual flux helps the individual flux is given by

$$L = \frac{L_1 L_2 - M^2}{L_1 + L_2 - 2M}$$

When mutual flux opposes the individual fluxes, the equivalent inductance is

$$L = \frac{L_1 L_2 - M^2}{L_1 + L_2 + 2M}$$

36. Magnetic Energy Stored in an Inductance

$$U_B = \int_0^i L i\, di = \frac{1}{2} L i^2$$

37. AC Circuits

Purely resistive circuit: When the current flowing through a pure resistance, no back *emf* is set-up and the applied voltage has to overcome the Ohmic drop of iR only. The instantaneous current is

$$i = \frac{V_m \sin \omega t}{R} = I_m \sin \omega t$$

The applied voltage and current are in phase with each other.

38. Periodic Time and Frequency

The time taken in second by an alternating quantity to complete one cycle is known as periodic time (T).

The number of cycles per second by an alternating quantity is known as frequency.

$$f = \frac{1}{T} \text{cycles/second or hertz}$$

39. Form Factor K_f

It is the ratio of RMS value to the average value of the wave. For a sine wave

$$\text{form factor } K_f = \frac{\text{RMS value}}{\text{Average value}}$$

40. Peak Factor K_j

It is the ratio of peak or maximum value to the RMS value of the wave

$$K_j = \frac{V_{\max}}{V_{\text{RMS}}}$$

41. Average Value of Sinusoidal Current

The average value of a sinusoidal wave is zero.

42. RMS Value of Alternating Current

$$I_{RMS} = \sqrt{\frac{1}{2\pi}\int i^2 dt}$$

$$= \sqrt{\frac{1}{2\pi}\int_0^{2\pi} I_m^2 \cdot \sin^2 \omega t \, d\omega t} = \frac{I_m}{\sqrt{2}}$$

43. Power Definitions and Relations

Power. The instantaneous power delivered is the product of instantaneous values of applied voltage and current $p = vi$

Power Factor: $pf = \cos\phi$, where ϕ is the phase angle between voltage and current. The power factor can never be greater than unity.

Apparent power	$P_a = V_{RMS} \, I_{RMS}$ (kVA)	
True power	$P = P_a \times$ power factor (kW)	
Reactive power,	$P_r = P_a \sin\phi$ (kVAR)	

where kVAR is kilo-amperes reactive.

44. Purely Inductive Circuit

It is observed that current lags behind the applied voltage by $\pi/2$ as shown in Fig. 12.11.

Instantaneous power, $p = vi$

Thus in purely inductive circuits, power absorbed is zero.

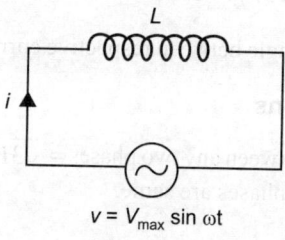

$v = V_{max} \sin \omega t$

Fig. 12.11

45. Purely Capacitive Circuit

When an alternating voltage $v = V_{max} \sin \omega t$ is applied across the capacitor of capacitance C, the instantaneous charge is $q = CV_{max} \sin \omega t$ and instantaneous current is given by

$$i = I_{max} \sin\left(\omega t + \frac{\pi}{2}\right)$$

It is observed that current leads the applied voltage by $\pi/2$ and capacitive reactance is $X_C = 1/\omega C$. This is shown in Fig. 12.12.

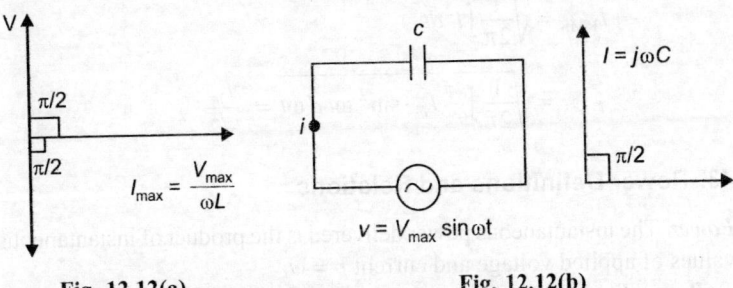

Fig. 12.12(a)   Fig. 12.12(b)

46. Mesh or Delta Connection

When the starting end of one coil is connected to the finishing end of another coil, delta connection is obtained (Fig. 12.13). In delta connection, line voltage V_L = phase voltage V_p, and

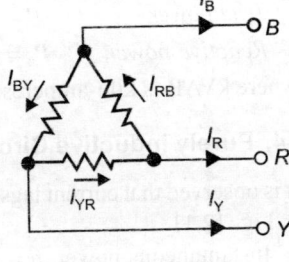

$$\text{line current } I_L = \sqrt{3} I_P;$$

$$\text{output power} = \sqrt{3} V_L I_L \cos\phi$$

Fig. 12.13

where ϕ is the phase angle between respective current and phase voltage.

47. Star Connections

Potential difference between any two phases = $\sqrt{3} V_P$. The current flowing through the lines and phases are same.

$$I_L = I_P$$

$$\text{Total power output} = \sqrt{3} V_L I_L \cos\phi$$

where ϕ is the angle between respective phase voltage and phase current.

48. Phase Sequence

Phase sequence is the order or sequence in which currents in different phases attain their maximum values one after the other. If the phase sequence is reversed by interchanging any two terminals of a three phase supply, the rotation of the motor would be reversed.

Power factor obtained by two Wattmeter method:

$$\cos \phi = \cos \left[\tan^{-1} \sqrt{3} \frac{(W_1 - W_2)}{W_1 + W_2} \right] \text{ for balanced load.}$$

49. Composite Magnetic Circuits

Consider a circular ring made of different magnetic materials of length l_1, l_2 and l_3, cross-sectional areas a_1, a_2 and a_3 and relative permeabilities μ_{r1}, μ_{r2}, μ_{r3} with an air gap as shown in Fig. 12.14.

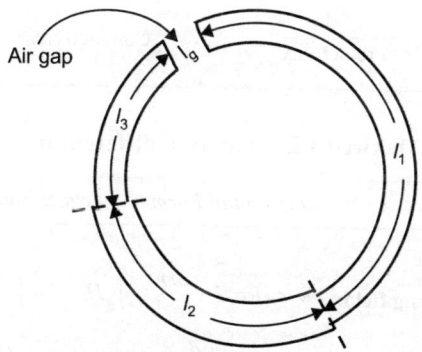

Fig. 12.14

Reluctance $\qquad S = \dfrac{l_1}{\mu_0 \mu_{r1} a_1} + \dfrac{l_2}{\mu_0 \mu_{r2} a_2} + \dfrac{l_3}{\mu_0 \mu_{r3} a_3} + \dfrac{l_g}{\mu_0 a}$

Total *mmf* = flux × reluctance

$$= \Phi \left[\frac{l_1}{\mu_0 \mu_{r1} a_1} + \frac{l_2}{\mu_0 \mu_{r2} a_2} + \frac{l_3}{\mu_0 \mu_{r3} a_3} + \frac{l_4}{\mu_0 a_g} \right]$$

and, total ampere-turns required = $H_1 l_1 + H_2 l_2 + H_3 l_3 + H_g l_g$

Comparison between Magnetic and Electric Circuits

Magnetic Circuit	Electric Circuit
Flux $= \dfrac{mmf}{\text{reluctance}}$	Current $= \dfrac{\text{emf}}{\text{resistance}}$
MMF (Amp. turns)	EMF (V)
Flux, Φ (Wb)	Current, I (A)
Flux density, $\boldsymbol{B}$ (Wb/m^2)	Current density, J (A/m^2)
Reluctance, $S = \dfrac{l}{\mu_0 \mu_r A}$	Resistance $R = \rho\dfrac{l}{A}$
Permeance $= \dfrac{1}{\text{reluctance}}$	Conductance $= \dfrac{1}{\text{resistance}}$
Reluctivity	Resistivity
Permeability $= \dfrac{1}{\text{reluctivity}}$	Conductivity $= \dfrac{1}{\text{resistivity}}$

Maxwell's Equations in different form

Description	Differential Form	Integral Form
(a) General case		
(i) **Time varying fields**	$\nabla \times \boldsymbol{H} = \boldsymbol{J} + \dfrac{\partial \boldsymbol{D}}{dt}$	$\oint_c \boldsymbol{H} \cdot dl = \int_s \boldsymbol{J} \cdot ds + \int_s \dfrac{\partial \boldsymbol{D}}{\partial t} \cdot ds$
	$\nabla \times \boldsymbol{E} = \dfrac{-\partial \boldsymbol{B}}{dt}$	$\oint_c \boldsymbol{E} \cdot dl = -\int \dfrac{\partial \boldsymbol{B}}{\partial t} \cdot ds$
	$\nabla \cdot \boldsymbol{D} = \rho$	$\oint_s \boldsymbol{D} \cdot ds = \int_v \rho \, dv$
	$\nabla \cdot \boldsymbol{B} = 0$	$\oint_s \boldsymbol{B} \cdot ds = 0$
	$\nabla \cdot \boldsymbol{J} = \dfrac{-\partial \rho}{\partial t}$	$\int_s \boldsymbol{J} \cdot ds = -\int_v \dfrac{\partial \rho}{\partial t} \cdot dV$
(ii) **Static fields**	$\nabla \times \boldsymbol{H} = \boldsymbol{J}$	$\oint_c \boldsymbol{H} \cdot dl = -\int_s \boldsymbol{J} \cdot ds$
	$\nabla \times \boldsymbol{E} = 0$	$\oint_c \boldsymbol{E} \cdot dl = 0$
	$\nabla \cdot \boldsymbol{D} = \rho$	$\oint_s \boldsymbol{D} \cdot ds = \int_v \rho \, dv$
	$\nabla \cdot \boldsymbol{B} = 0$	$\oint_s \boldsymbol{B} \cdot ds = 0$
	$\nabla \cdot \boldsymbol{J} = 0$	$\int_s \boldsymbol{J} \cdot ds = 0$

Description	Differential Form	Integral Form
(b) Perfect dielectrics $\rho = 0 \quad J = 0$		
(i) **Time varying fields**	$\nabla \times H = \dfrac{\partial D}{\partial t}$	$\oint_c H \cdot dl = \int_s \dfrac{\partial D}{\partial t} \cdot ds$
	$\nabla \times E = \dfrac{-\partial B}{\partial t}$	$\oint_c E \cdot dl = -\int_s \dfrac{\partial B}{\partial t} \cdot ds$
	$\nabla \cdot D = 0$	$\oint_s D \cdot ds = \int_v \rho \, dv$
	$\nabla \cdot B = 0$	$\oint_s B \cdot ds = 0$
(ii) **Static fields**	$\nabla \cdot H = 0$	$\oint_c H \cdot dl = 0$
	$\nabla \times E = 0$	$\oint_c E \cdot dl = 0$
	$\nabla \cdot D = 0$	$\oint_s D \cdot ds = \int_v \rho \, dv$
	$\nabla \cdot B = 0$	$\oint_s B \cdot ds = 0$
(c) Good Conductors $J > \dfrac{\partial D}{\partial t}, \rho = 0$		
(i) **Time varying fields**	$\nabla \times H = J$	$\oint_c H \cdot dl = \int_s J \cdot ds + \int_s \dfrac{\partial D}{\partial t} \cdot ds$
	$\nabla \times E = \dfrac{-\partial B}{dt}$	$\oint_c E \cdot dl = -\int_s \dfrac{\partial B}{\partial t} \cdot ds$
	$\nabla \cdot D = 0$	$\oint_s D \cdot ds = \int_v \rho \, dv$
	$\nabla \cdot B = 0$	$\oint_s B \cdot ds = 0$
	$\nabla \cdot J = 0$	$\oint_s J \cdot ds = -\dfrac{d}{dt}\int \rho dv$
(ii) **Static fields**	$\nabla \times H = J$	$\oint_c H \cdot dl = \int_s J \cdot ds$
	$\nabla \times E = 0$	$\oint_c E \cdot dl = 0$
	$\nabla \cdot D = 0$	$\oint_s D \cdot ds = \int_v \rho \, dv$
	$\nabla \cdot B = 0$	$\oint_s B \cdot ds = 0$
	$\nabla \cdot J = 0$	$\int_s J \cdot ds = 0$

Chapter 13

Circuit Theory and Network

1. Electric Current and Voltage

Electric current can be expressed as

$$i = \frac{dq}{dt}$$

The unit of current is ampere (A); an ampere is 1 coulomb per second.
Current is the time rate of flow of electric charge. *Charge* is the
quantity of electricity responsible for electric phenomena.

$$q = \int_0^t i \, d\tau + q(0)$$

The *voltage* across an element is the work required to move a positive
charge of 1 coulomb from the first terminal through the element to the
second terminal. The unit of voltage is volt (V).

$$v = \frac{dw}{dq}$$

where v is voltage, w is energy, and q is charge. A charge of 1 coulomb
delivers an energy of 1 joule as it moves through a voltage of 1 volt.

Power is the time of expanding or absorbing energy. Thus, the equation

$$p = \frac{dw}{dt}$$

where p is the power in watts, w is energy in joules, and t is the time in
seconds;

$$p = v \cdot i$$

2. Current Flow in Circuit Element

When energy is delivered to the element, the voltage drop across two terminals a and b is said to be a voltage v as shown in Fig. 13.1.

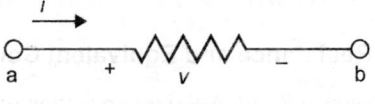

Fig. 13.1

A *passive element* absorbs energy

$$w = \int_{-\infty}^{t} v\, i\, d\tau \geq 0$$

when both v and i are of the same sign.

3. Resistance and Ohm's Law

Resistance is the physical property of an element or device that opposes the flow of current; it is represented by the symbol R. Resistance R is defined as

$$R = \frac{\rho L}{A}$$

where A is the cross-sectional area, ρ is the resistivity, L is the length of the wire element.

Ohm's law relates the voltage and current in a resistance given by

$$v = Ri$$

The unit of resistance R is ohm and is usually abbreviated by the symbol Ω, where $1\ \Omega = 1$ V/A.

Ohm's law can also be written as $i = Gv$
where G denotes the conductance in siemens (S).

The power delivered to a resistor is

$$p = vi = \frac{v^2}{R} = i^2 R$$

4. Kirchhoff's Laws

Kirchhoff's Current Law (KCL): The algebraic sum of the currents at a node at any instant is zero.

$$\sum_{n=1}^{N} i_n = 0$$

Kirchhoff's Voltage Law (KVL): The algebraic sum of voltage around any closed path in a circuit is identically zero for all time.

$$\sum_{n=1}^{N} v_n = 0$$

5. Equivalent Resistance and Equivalent Conductance

An equivalent resistance R_s, for a series connection of N resistors is

$$R_s = \sum_{j=1}^{N} R_j$$

An equivalent conductance, G_p for a parallel connection of N conductances is

$$G_p = \sum_{j=1}^{N} G_j$$

6. Voltage and Current Divider Circuits

The voltage v_n, across the nth resistor of N resistors connected in a series is

$$v_n = \frac{R_n}{R_1 + R_2 + ... + R_N} v_s = \frac{R_n}{\sum_{j=1}^{N} R_j} v_s$$

where v_s is the source voltage connected in series with the resistors.

The current i_n in the conductance G_n connected in a parallel set of N conductances is

$$i_n = \frac{G_n i_s}{\sum_{j=1}^{N} G_j}$$

where i_s is a source current connected in parallel with the parallel set of conductance.

7. Node Voltages

The node voltage matrix equation for a circuit with N unknown node voltages is

$$G v = i_s$$

where

$$v = \begin{bmatrix} v_a \\ v_b \\ . \\ . \\ . \\ v_N \end{bmatrix}$$

which is the vector of unknown node voltages. The matrix

$$i_s = \begin{bmatrix} i_{s1} \\ i_{s2} \\ . \\ . \\ . \\ i_{sN} \end{bmatrix}$$

is the vector consisting of the N current sources, where i_{sN} is the sum of all the source currents into the node n.

When there are no dependent sources within the circuit, the conductance matrix is symmetric as

$$G = \begin{bmatrix} \sum_a G & -G_{ab} & \cdots & -G_{aN} \\ -G_{ab} & \sum_b G & \cdots & -G_{bN} \\ \vdots & & & \\ -G_{aN} & -G_{bN} & \cdots & \sum_N G \end{bmatrix}$$

where $\sum_n G$ is the sum of the conductances at node n and G_{ij} is the sum of the conductances at connecting nodes i and j. When the circuit includes dependent sources, the G matrix is not symmetric.

8. Mesh Current Analysis

Assume a planar network with N meshes containing N mesh currents flowing clockwise. The matrix equation for mesh current analysis with no dependent sources is

$$\mathbf{Ri} = v_s$$

where R is a symmetric matrix with a diagonal consisting of the sum of resistances in each mesh, and the off-diagonal elements are the negative of the resistances connecting two meshes. The matrix i consists of the mesh currents as

$$i = \begin{bmatrix} i_1 \\ i_2 \\ \cdot \\ \cdot \\ \cdot \\ i_N \end{bmatrix}$$

For N mesh current, the source matrix v_s is

$$v_s = \begin{bmatrix} v_{s1} \\ v_{s2} \\ \cdot \\ \cdot \\ \cdot \\ v_{sN} \end{bmatrix}$$

where v_{sj} is the sum of the sources in the jth mesh with the appropriate sign assigned to each source.

When dependent sources are present within the circuit, the R matrix is not symmetric.

9. Voltage and Current Source Transformations

A *source transformation* is a procedure for transforming one source into another while retaining the terminal characteristics of the original source. The transformation of a voltage source in series with a resistance R_s is transformed into a current source with a resistance R_p in parallel is shown in Fig. 13.2a.

The transformation of a current source in parallel with a resistance R_p can be transformed into a voltage source in series with a resistance R_s as shown in Fig. 13.2b.

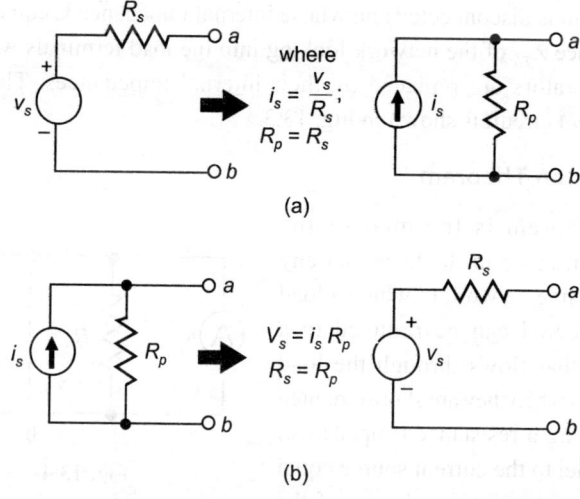

(a)

$V_s = i_s R_p$
$R_s = R_p$

(b)

Fig. 13.2

10. The Superposition Principle

In a linear circuit containing independent sources, the voltage across (or the current through) any element may be obtained by adding algebraically all the individual voltages (or currents) caused by each independent source acting alone, with all other independent voltage sources replaced by short circuits and all other independent current sources replaced by open circuits.

The voltage across an element v

$$v = \sum_{j=1}^{N} v_j$$

where v_j is the voltage due to the jth source with all other sources disabled.

11. Thevenin Theorem

Thevenin theorem states that for any linear active network to which a load is connected can be reduced to a voltage generator whose generated voltage E_{Th} is equal to the open circuit voltage that appears across the load terminal

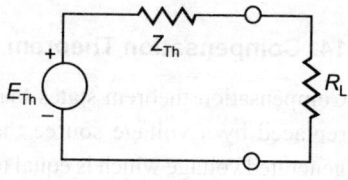

Fig. 13.3

when load is disconnected and whose internal impedance is equal to the impedance Z_{Th} of the network looking into the load terminals when all the generators are replaced by their internal impedances. Thevenin equivalent circuit is shown in Fig. 13.3.

12. Norton Theorem

This theorem is the dual of the Thevenin theorem. It states that any linear active network to which a load is connected can be reduced to a current that flows through the load terminals when they are short-circuited and having a resistance (impedance) in parallel to the current source equal to the resistance (impedance) of the network (Fig. 13.4).

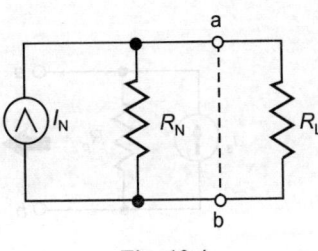

Fig. 13.4

13. Tellegan Theorem

It states that in an arbitrarily lumped network subject to KVL and KCL constraints, the product of all branch currents and branch voltages must be zero. Tellegen's theorem may be summarized by the equation

$$\sum_{k=1}^{b} v_k j_k = 0$$

where the lower case letters v and j represent instantaneous values of the branch voltages and branch currents, respectively, and b is the total number of branches. A matrix representation employing the branch current and branch voltage vectors also exists. Because V and J are column vectors, Thus,

$$V \cdot J = V^T J = J^T V$$

14. Compensation Theorem

Compensation theorem states that any resistance in a network may be replaced by a voltage source that has zero internal resistance and a generated voltage which is equal to the potential drop across the replaced resistance by current flowing through it and directed against the current.

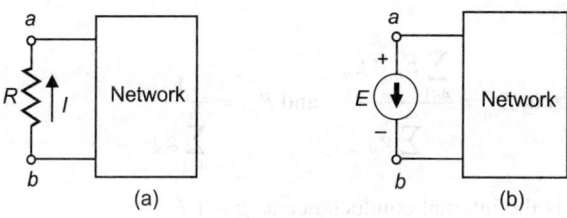

Fig. 13.5

In Fig. 13.5a, the resistance R through which current I flows can be replaced by a voltage source E [Fig. 13.5b] whose generated voltage is $E = I \times R$ and directed against the direction of current I.

15. Maximum Power Transfer Theorem

This theorem states that the maximum power delivered by a source represented by its Thevenin equivalent circuit is attained when the load R_L is equal to the Thevenin resistance R_T (Fig. 13.6).

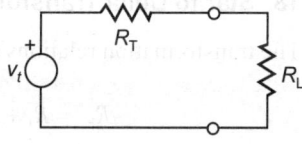

Fig. 13.6

16. Parallel Generator Theorem

The several voltage generators connected in parallel can be replaced by a single generator of equivalent voltage and equivalent internal resistances as depicted in Fig. 13.7.

$$E_{eq} = \frac{E_1 g_1 - E_2 g_2 + E_3 g_3}{g_1 + g_2 + g_3}$$

and

$$R_{eq} = \frac{1}{g_1 + g_2 + g_3}$$

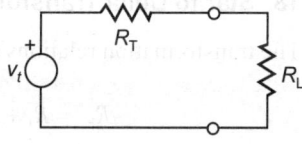

Fig. 13.7

Generalizing $E_{eq} = \dfrac{\sum\limits_{n=1}^{\infty} E_n \times g_n}{\sum\limits_{n=1}^{\infty} g_n}$, and $R_{eq} = \dfrac{1}{\sum\limits_{n=1}^{\infty} g_n}$

where g is the internal conductance as $g = 1/R$

17. Efficiency of Power Transfer

The *efficiency of power transfer* is defined as the ratio of the power delivered to the load P_{out}, to the power supplied by the source P_{in}.

$$\eta = P_{out}/P_{in}$$

18. Star to Delta Transformation

The transformation relations are as given below (Fig. 13.8):

$$R_{12} = R_1 + R_2 + \frac{R_1 R_2}{R_3}$$

$$R_{23} = R_2 + R_3 + \frac{R_2 R_3}{R_1}$$

$$R_{31} = R_3 + R_1 + \frac{R_3 R_1}{R_2}$$

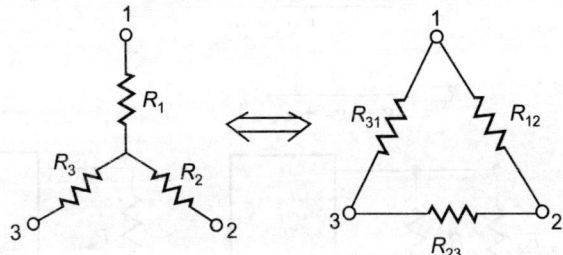

Fig. 13.8

19. Delta to Star Transformation

The transformation relations are as follows:

$$R_1 = \frac{R_{12} \cdot R_{31}}{R_{12} + R_{23} + R_{31}}$$

$$R_2 = \frac{R_{23} \cdot R_{12}}{R_{12} + R_{23} + R_{31}}$$

$$R_3 = \frac{R_{23} \cdot R_{31}}{R_{12} + R_{23} + R_{31}}$$

20. Circuits with Energy Storage Elements

(i) *Capacitors*: *Capacitance* is a measure of the ability of a device to store energy in the form of (separated) charge or in the form of an electric field:

$$q = Cv$$

where q is the charge, v is the voltage across the element, and C is the capacitance measured in farad (F).

The current through a capacitor is

$$i = C\frac{dv}{dt}$$

The voltage across a capacitor C is

$$v = \frac{1}{C}\int_{t_0}^{t} i\,d\tau + v(t_0)$$

where $v(t_0)$ is the voltage at time t_0.

(ii) *Inductors*: *Inductance* is a measure of the ability of a device to store energy in the form of a magnetic field. The voltage across an indicator is

$$v = L\frac{di}{dt}$$

where i is the current through the inductor and L is the inductance measured in henry (H).

The current in an inductor is

$$i = \frac{1}{L}\int_{t_0}^{t} v\,d\tau + i(t_0)$$

(iii) *Energy Stored in Inductors and Capacitors*

$$\text{Energy stored in capacitor} = \frac{1}{2}Cv^2$$

and

$$\text{Energy stored in inductor} = \frac{1}{2}Li^2$$

(iv) **Series and Parallel Inductors**: A series connection of N inductors can be represented by one series equivalent inductor L,

Thus, $\qquad L_s = L_1 + L_2 + L_3 + ... + L_n$ or $L_s = \displaystyle\sum_{n=1}^{N} L_n$

A parallel connection of N inductors can be represented by one equivalent inductor L_p:

$$\frac{1}{L_p} = \frac{1}{L_1} + \frac{1}{L_2} + \frac{1}{L_3} + ... + \frac{1}{L_n} \text{ or } \frac{1}{L_p} = \sum_{n=1}^{N} \frac{1}{L_n}$$

(v) **Series and Parallel Capacitors**: The equivalent capacitance of a set of N parallel capacitors is simply the sum of the individual capacitances:

$$C_p = C_1 + C_2 + C_3 + ... + C_n \text{ or } C_p = \sum_{n=1}^{N} C_n$$

A series connection of N capacitors can be represented by one equivalent capacitance C_s:

$$\frac{1}{C_s} = \frac{1}{C_1} + \frac{1}{C_2} + \frac{1}{C_3} + ... + \frac{1}{C_n} \text{ or } \frac{1}{C_s} = \sum_{n=1}^{N} \frac{1}{C_n}$$

(vi) **The Natural Response of an RL or RC Circuit**: The *natural response* of a circuit depends only on the internal energy storage of the circuit and not on external sources. The natural response of a series connection of a resistor R and a capacitor C is

$$v = V_0 e^{-t/RC}$$

where $v(0) = V_0$ is the initial voltage on the capacitor and v is the capacitor voltage.

The natural response of a series connection of a resistor R and inductor L is

$$i = I_0 \cdot e^{-Rt/L}$$

where $i(0) = I_0$ is the initial current and i is the inductor current.

(vii) **The Forced Response of an RL or RC Circuit Excited by a Constant Source**: The *forced response* of a circuit is the behaviour exhibited in reaction to one or more independent signal source. The forced response of an *RC* circuit is

$$v(t) = v(\infty) + [v(0) - v(\infty)]e^{-t/RC}$$

where $v(\infty)$ is the steady-state value at $t = \infty$.

The forced response of an *RL* circuit is

$$i(t) = i(\infty) = [i(0 - i(\infty))]\ e^{-t/\tau}$$

where $\tau = L/R$.

(viii) ***The Natural Response of an RLC Circuit***: The differential equation for a parallel connection of an R, L and C

$$= \frac{d^2v}{dt^2} + \frac{1}{RC}\frac{dv}{dt} + \frac{v}{LC} = 0$$

where v is the voltage across the capacitor (Fig. 13.9).

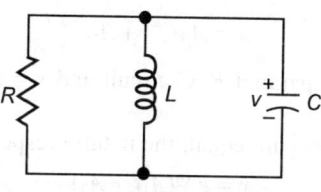

Fig. 13.9

The differential equation for the series connection of R, L and C

$$= \frac{d^2i}{dt^2} + \frac{R}{L}\frac{di}{dt} + \frac{i}{LC} = 0$$

where i is the current through the inductor (Fig. 13.10)

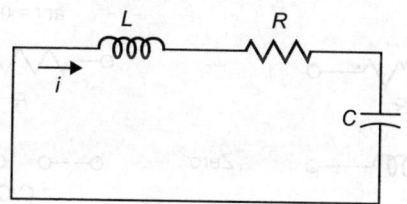

Fig. 13.10

The characteristic equation is

$$s^2 + a_1 s + a_0 = 0$$

or

$$s^2 + 2\alpha s + \omega_0{}^2 = 0$$

Then the roots of the characteristic equation are

$$s_1 = -\alpha + \sqrt{\alpha^2 - \omega_0^2}$$

$$s_2 = -\alpha - \sqrt{\alpha^2 - \omega_0^2}$$

where $\omega_0 = 1/\sqrt{LC}$ is called the *resonant frequency*.

The roots of the characteristic equation assume three possible conditions.
1. Two real and distinct roots when $\alpha^2 > \omega_0^2$
2. Two real equal roots when $\alpha^2 = \omega_0^2$
3. Two complex roots when $\alpha^2 < \omega_0^2$

When the two roots are real and distinct, the circuit is said to be *overdamped*. When the roots are both real and equal, the circuit is *critically damped*. When the two roots are complex conjugates, the circuit is said to be *underdamped*.

The overdamped natural response is

$$x = A_1 e^{-s_1 t} + A_2 e^{-s_2 t}$$

where $x = v$ for the parallel *RLC* circuit and $x = i$ for the series *RLC* circuit.

When the two roots are equal, the natural response is

$$x = e^{-\alpha t}(A_1 t + A_2)$$

When the circuit is underdamped, we have

$$x = e^{-\alpha t}(B_1 \cos \omega_d t + B_2 \sin \omega_d t)$$

where $\omega_d = \sqrt{\omega_0^2 - \alpha^2}$, the damped resonant frequency.

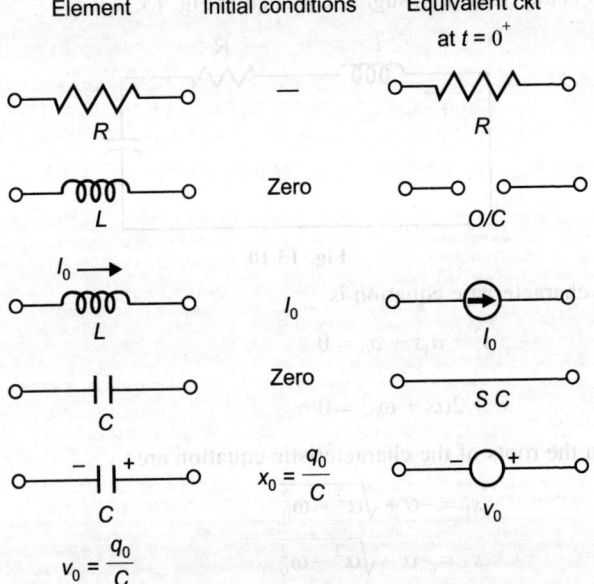

Fig. 13.11

21. Initial Conditions

Initial conditions in various elements are shown in Fig. 13.11.

22. Final Conditions in Elements Behaviour

The equivalent circuits of the elements for the final condition (*i.e.* at $t = \alpha$) are shown in Fig. 13.12.

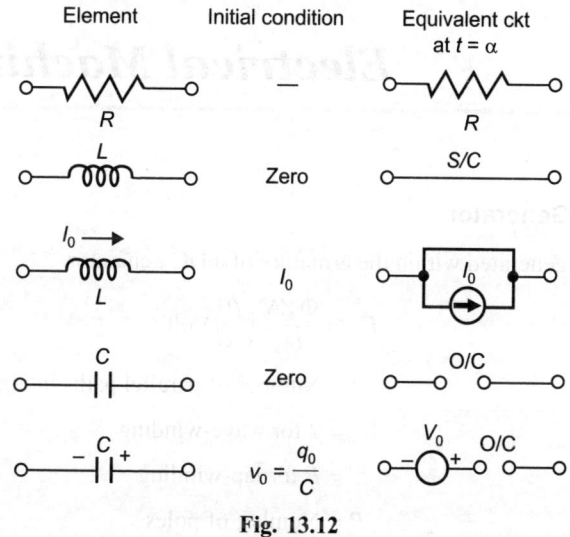

Fig. 13.12

Chapter 14

Electrical Machines

1. DC Generator

The emf generated within the armature of a DC generator.

$$E_g = \frac{\Phi ZN}{60}\left(\frac{P}{A}\right)\text{Volt}$$

where

A = Number of parallel paths in armature

= 2 for wave-winding

= P for lap-winding

P = Number of poles

N = Speed in RPM

Z = Number of conductors in the armature

Φ = Flux/pole in Wb

Condition for Maximum Efficiency:

Armature copper loss = Iron loss

The load current corresponding to maximum efficiency is given by

$$I = \sqrt{\frac{W_{Cu}}{R_a}}$$

where W_{Cu} is the copper losses and R_a is armature resistance.

Demagnetizing amp-turns per Pole: $AT_d/\text{pole} = Z\, I_c\, \dfrac{\theta}{360}$

Cross-magnetizing amp-turns per pole:

$$AT_c/\text{pole} = ZI_c\left[\frac{1}{2P} - \frac{\theta}{360}\right],$$

where Z is the total number of conductors, I_c is the current per conductor and θ is the angle of lead of brush from GNP in mechanical degrees and P the number of poles.

    Reactance Voltage: The self-induced emf in the coil undergoing commutation is known as *reactance voltage.*

$$\text{Value of reactance voltage} = 1.11\ L(2I_c/T_c)$$

where T_c is the commutation time, and L is the self-inductance of the coil.

$$\text{Amp.-turns/pole for compensating winding} = \frac{ZI_c}{2P} \times \frac{\text{Pole arc}}{\text{Pole pitch}}$$

2. Types of DC Generators

I Series Wound

(i) $I_a = I_{se} = I_L = I$

(ii) $V = E_g - I(R_a + R_{se})$

(iii) Power developed $P_{dev} = E_g I$

(iv) Power delivered, $P_{dil} = VI$

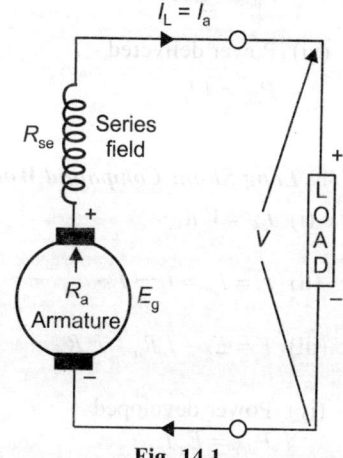

Fig. 14.1

II Shunt Wound

(i) $I_{sh} = V/R_{sh}$

(ii) $I_a = I_L + I_{sh}$

(iii) $V = E_g - I_a R_a$

(iv) Power developed,
    $P_{dev} = E_g I_a$

(v) Power delivered,
    $P_{dil} = VI_L$

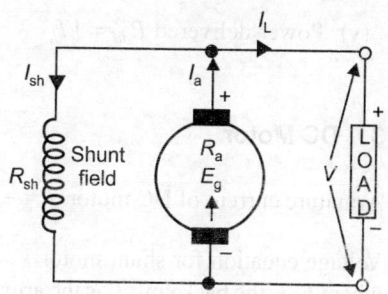

Fig. 14.2

III Compound Wound

(a) Short shunt compound wound (b) Long shunt compound wound.

(a) *Short Shunt Compound Wound*

(i) $I_{se} = I_L$

(ii) $I_{sh} = \dfrac{V + I_{se}R_{se}}{R_{sh}}$

(iii) $I_a = I_L + I_{sh}$

(iv) $V = E_g - I_aR_a - I_{se}R_{se}$

(v) Power developed
$$P_{dev} = E_g I_a$$

(vi) Power delivered
$$P_{dil} = VI_L$$

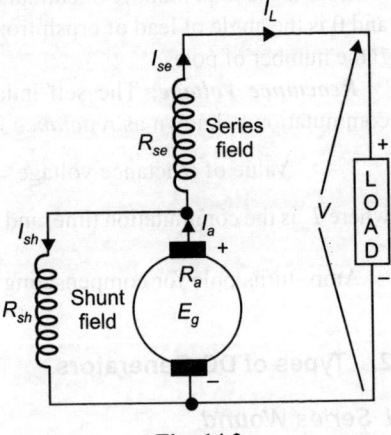

Fig. 14.3

(b) *Long Shunt Compound Wound*

(i) $I_{sh} = V/R_{sh}$

(ii) $I_a = I_{se} = I_L + I_{sh}$

(iii) $V = E_g - I_aR_a - I_{se}R_{se}$

(iv) Power developed
$$P_{dev} = E_g I_a$$

(v) Power delivered $P_{dil} = VI_L$

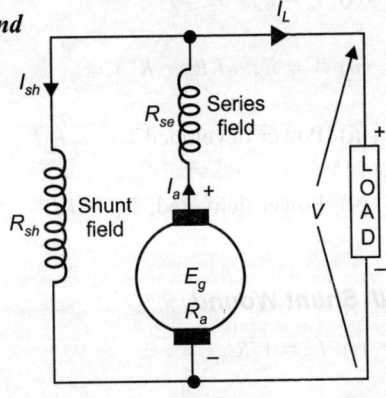

Fig. 14.4

3. DC Motor

Armature current of DC motor $I = \dfrac{V - E_b}{R_a}$

Voltage equation for shunt motor $V = E_b + I_aR_a + $ brush drop
where E_b is the back emf, I_a is the armature current, and R_a is the armature
resistance, and V is the supply voltage.

Power developed in DC motor, $E_b I_a = V I_a - I_a^2 R_a$

Condition for Maximum Power Developed in Armature of DC Motor:
$E_b = V/2$

Armature Torque of a DC Motor:
$$T_a = 0.159 \Phi Z I_a \,(P/A) \text{ Nm}$$
$$= 0.0162 \Phi Z I_a \,(P/A) \text{ kg} \cdot \text{m}$$

$$\textit{Useful torque} = \frac{\text{Output is watts}}{2\pi N} \text{ Nm}$$

Speed of DC Motor:

$$N = \frac{E_b}{\Phi}\left(\frac{A}{ZP}\right) \text{ rev. per sec.}$$

$$\% \text{ speed regulation} = \frac{N_0 - N_f}{N_f} \times 100$$

where N_0 and N_f are the no load and full load speeds.

Speed Control of DC Motors:

$$\textbf{The Speed of DC Motor, } N = \frac{V - I_a R_a}{\Phi} \cdot \frac{60 A}{ZP}$$

4. Types of DC Motors

I Series Wound Motor

(i) Back emf
$E_b = V - I(R_a + R_{se})$

(ii) $I_a = I_{se} = I$

(iii) Power Input $= VI$

(iv) Power developed
$P_{dev} = E_b I = VI - I^2(R_a + R_{se})$

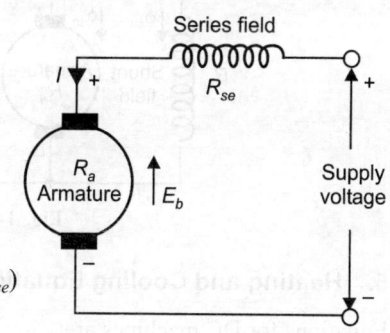

Fig. 14.5

II Shunt Wound Motor

(i) $I_L = I_a + I_{sh}$

(ii) $I_{sh} = V/R_{sh}$

(iii) Back emf $E_b = V - I_a R_a$

(iv) Power Input $= VI_L$

(v) Power developed $P_{\text{dev}} = VI_L - VI_{sh} - I_a^2 R_a$

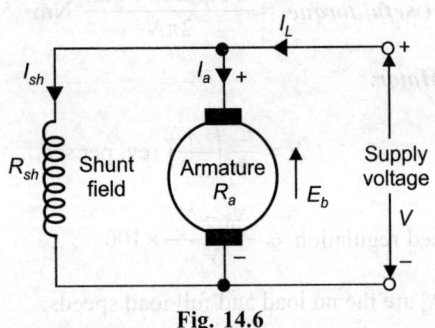

Fig. 14.6

III Compound Wound Motors

(a) Cummulative compound $\phi = \phi_{sh} + \phi_{se}$ (b) Differential compound

$$\phi = \phi_{se} - \phi_{sh}$$

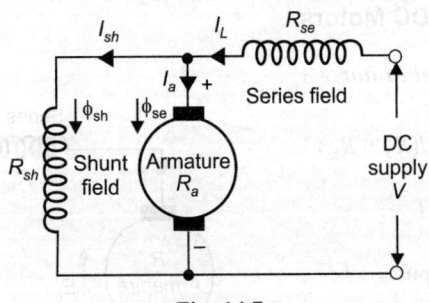

Fig. 14.7

5. Heating and Cooling Equations

Equations for DC machines are:

$$\theta = \theta_m \, (1 - e^{-t/T_h}) \qquad \text{...(Heating)}$$

$$\theta = \theta_i \, (1 - e^{-t/T_c}) \qquad \text{...(Cooling)}$$

where θ_m is the final temperature rise while heating, θ_i is the initial temperature rise over ambient medium $°C$, T_h and T_c are the heating and cooling time constants, t is the time (sec).

6. Transformers

EMF Equation: Induced e.m.f in secondary winding of transformers

$$E_2/phase = 4.44 \; \Phi_m f N_2 \text{ volts}$$

where Φ_m is the maximum value of flux in webers, N_1, N_2 being the number of turns on the primary and secondary winding and f the supply frequency in Hz.

Voltage Transformation Ratio: In an ideal transformer

$$\frac{V_2}{V_1} = \frac{I_1}{I_2} = \frac{N_2}{N_1} = K$$

Equivalent Resistance and Reactance of Transformer:

Total resistance of transformer referred to primary, $R_{01} = r_1 + r_2/K^2$

Total reactance of transformer referred to primary, $X_{01} = x_1 + x_2/K^2$

Total resistance of transformer referred to secondary, $R_{02} = r_2 + K^2 r_1$

Total reactance of transformer referred to secondary, $X_{02} = x_2 + K^2 x_1$

Transformer Efficiency:

$$\eta = \frac{x \cdot V_2 I_2 \cos\phi}{V_2 I_2 \cos\phi + P_i + x^2 P_c}$$

where V_2 is the secondary terminal voltage, I_2 secondary current at load, $\cos\phi$ is the power factor of load and x is the fraction of the load, P_i and P_{cu} being the iron and full load copper losses.

Percentage Voltage Regulation $= \left(\dfrac{I_2 R_{02} \cos\phi \pm I_2 X_{02} \sin\phi}{E_2} \right) \times 100$

where R_{02} and X_{02} being the transformer resistance and reactance referred to the secondary.

Moreover, + sign to be used for inductive load, and – sign to be used for capacitive load.

Condition for Maximum Efficiency

Iron loss, $P_i = x^2 P_{cv}$ = variable copper losses

Hysteresis loss, $\quad W_h = \eta B_m^{1.6} f V$ Watts

Eddy current loss, $\quad W_e = P B_m^2 f^2 t^2$ Watts

where V and t being the volume and thickness of the core laminations.

Output Current at which Maximum Efficiency occurs:

$$= \text{Full load current} \sqrt{\frac{\text{Iron loss}}{\text{FL copper loss}}}$$

$$\textbf{All day Efficiency} = \frac{\text{Output in kWh}}{\text{Input in kWh}} \text{ (for 24 h)}$$

$$= \frac{\text{Output (kWhr)}}{\text{Output (kWhr)} + \text{Copper losses (kWhr)} + \text{Iron loss} \times 24 \,\text{h}}$$

Load Sharing by Two Transformers A and B:

$$P_A = \frac{Z_B}{Z_A + Z_B} \cdot P$$

where Z_A and Z_B being the impedance of two transformers A and B and P is the total load to be shared by them.

7. Induction Motor

Synchronous speed $\quad N_s = \dfrac{120 f}{P}$

where f and p being the supply frequency and number of poles.

Slip of an Induction Motor

$$s = \frac{\text{synchronous speed} - \text{rotor speed}}{\text{synchronous speed}}$$

Rotor current frequency, $f' = \text{slip} \times \text{supply frequency} = s \times f$

Condition for Maximum Starting Torque: Rotor resistance per phase = rotor *reactance, i.e.* $R_2 = X_2$.

Running Torque: The torque under running conditions is maximum at that value of slip s, which makes rotor resistance per phase equal to slip times the rotor reactance per phase, or $R_2 = s X_2$

$$\text{Torque } T = \frac{k s R_2 E_2^2}{R_2^2 + s^2 \cdot X_2^2}$$

The full load torque T_f and maximum torque T_{max} are related by:

$$\frac{T_f}{T_{max}} = \frac{2as}{a^2 + s^2}$$

where

$$a = \frac{R_2}{X_2}$$

Rotor copper loss in induction motor $= s \times$ Rotor input

The electrical equivalent of mechanical load on motor,

$$R_L = R_2 \left(1/s - 1\right)$$

Slip corresponding to maximum gross power output

$$s_m = \frac{R_2 / K^2}{R_2 / K^2 + Z_{01}}$$

Maximum gross power output:

$$P_{g\,max} = \frac{3V_1^2}{2\left(R_{01} + Z_{01}\right)} \quad \text{where, } V_1 \text{ is the input voltage}$$

Gross torque $T_g = \dfrac{\text{Rotor output in watts}}{2\pi N}$

Rotor copper loss $= T_g \times 2\pi \left(N_s - N\right)$

$$\frac{\text{Rotor gross output}}{\text{Rotor input}} = \left(1 - s\right) = \frac{N}{N_s}$$

Rotor efficiency $= \dfrac{N}{N_s}$

$$\frac{\text{Rotor copper loss}}{\text{Rotor gross output}} = \frac{s}{1 - s}$$

Speed of cummulative cascade set $N_{sc} = \dfrac{120f}{P_a + P_b}$

where P_a and P_b are the numbers of poles of two machines.

9. Methods of Starting of Induction Motor

(i) *Torque Developed by Direct Switching at Starting*:

$$T_{st} = T_f \left(\frac{I_{st}}{I_f}\right)^2 \cdot s_f$$

where I_{st} and I_f being the starting and full load currents, T_f and s_f are the full load torque and slip. Starting current is equal to the short-circuit current.

(ii) *Torque Developed by Reduced Voltage Method*:

$$T_{st} = T_f \cdot k^2 \left(\frac{I_{st}}{I_f} \right)^2 \cdot s_f$$

where k is the factor by which voltage is reduced using resistance starter. Starting current $= k I_{sc}$, where I_{sc} is the short-circuit current.

(iii) *Auto-Transformer Starting Method*

$$T_{st} = T_f \left(\frac{I_{st}}{I_{sc}} \right)^2 s_f = T_f k^2 \left(\frac{I_{st}}{I_f} \right)^2 s_f$$

$$= k^2 \times \text{torque obtained by DOL}$$

where k is the transformation ratio; supply current $= k^2 I_{sc}$

(iv) *Star-Delta Starting Method*:

$$T_{st} = \frac{1}{3} T_f \left(\frac{I_{sc}}{I_f} \right)^2 \cdot s_f$$

Here starting torque is reduced to one third of DOL torque and starting line current is reduced to one-third of line current with DOL starting.

10. Alternator

The frequency of alternating current, $f = \dfrac{PN}{120} \, \text{Hz}$

The emf equation of an alternator,

$$E = 4.44 \, K_c \, K_d \, \Phi f T \text{ volts/phase}$$

where K_c is coil span factor, $(K_c = \cos \alpha/2)$, α is the angle by which the coil span falls short and

K_d is the distribution factor and is equal to $\dfrac{\sin m\beta/2}{m\sin \beta/2}$

n = number of slots/pole/phase, $\beta = \dfrac{180°}{\text{No. of slots/pole}}$

Voltage Regulation:

The terminal voltage V is less than open-circuit voltage E_a because of

  (i) armature voltage drops IR_a

 (ii) synchronous reactance voltage drops IX_s, and

(iii) armature reaction voltage drops $I_a X_a$

    Therefore, $\qquad E_a = V + IR_a + IX_s + IX_a$

$$\% \textbf{ Voltage Regulation} = \frac{E_0 - V}{V} \times 100$$

Synchronizing current $I_{sy} = \dfrac{E_r}{Z_s}$

Synchronizing power $P_{sy} = \alpha E I_{sy} \approx \dfrac{\alpha E^2}{X_s}$ watt/phase

Load angle $\alpha = \dfrac{P}{2} \times$ angle of retardation in mechanical degree

Synchronizing torque $T_{sy} = \dfrac{60 \times P_{sy}}{2\pi N_s}$ Nm

The mechanical power developed

$$P_m = \frac{E_b V}{Z_s} \cos(\theta - \alpha) - \frac{E_b^2}{Z_s} \cos\theta$$

where α = load angle, and θ is the internal angle of the motor for a constant supply voltage V and induced *emf* E_b.

Maximum power developed, $P_m = \dfrac{E_b V}{Z_s} - \dfrac{E_b^2}{Z_s^2}$

Maximum power developed, $P_m = \dfrac{V^2}{4R_a}$

Induced emf when pf is lagging, $E_b = \sqrt{V^2 + E_r^2 - 2VE_r \cos(\theta - \phi)}$

Induced emf when pf is leading

$$E_b = \sqrt{V^2 + E_r^2 - 2VE_r \cos(\theta + \phi)}$$

Also $\qquad \dfrac{E_r}{\sin\alpha} = \dfrac{E_b}{\sin(\theta + \phi)}$

11. Converters and Rectifiers

(i) $\dfrac{E_p}{E_{DC}} = \dfrac{\sin(\pi/m)}{\sqrt{2}}$

where E_p is RMS phase voltage and m is the number of phases.

(ii) $I_p = \dfrac{\sqrt{2}\,I_{DC}}{\eta m \sin(\pi/m) \cdot \cos\phi}$ where, I_p is the RMS phase current

$= \dfrac{\sqrt{2}\,I_{DC}}{m \sin(\pi/m)}$, assuming upf and 100% efficiency

(iii) $I_{sr} = \dfrac{2.83\,I_{DC}}{m}$, Assuming upf and 100% efficiency

where, I_{sr} is the load current in general, for m-phase rotary converter

Voltage Relation

$$E_{DC} = \sqrt{2}\,E_{AC}\,\frac{m}{\pi}\cdot\sin(\pi/m)$$

Current Relation

$$I_{rms} = I_{DC}/\sqrt{m}$$

where, E_{DC} = Average value of no load DC voltage,

E_{AC} = rms value of secondary phase neutral voltage

I_{RMS} = Value of secondary current

Chapter 15

Instrumentation

Current and Potential Transformers (CT and PT)

1. Normal Ratio

$$K_n = \frac{\text{Rated primary current}}{\text{Rated secondary current}}; \text{ for a CT}$$

$$= \frac{\text{Rated primary voltage}}{\text{Rated secondary voltage}}; \text{ for a PT}$$

2. Turns Ratio (n)

$$= \frac{\text{Number of turns of secondary winding}}{\text{Number of turns of primary winding}}$$

3. Transformation Ratio

For a current transformer

$$R \cong n + \frac{I_0}{I_s} \sin(\delta + \alpha)$$

where, n = turns ratio, I_0 = exciting current
I_s = secondary current
δ = angle between secondary induced emf and secondary current, and
α = angle between exciting current I_0 and working flux.

4. Phase Angle for Current Transformer

$$\theta = \frac{180}{\pi}\left[\frac{I_m \cos\delta - I_e \sin\delta}{nI_s}\right] \text{ degrees}$$

125

5. Ratio Error $= \left(\dfrac{\text{nominal ratio} - \text{actual ratio}}{\text{actual ratio}} \right) \times 100\% = \dfrac{K_n - R}{R} \times 100$

6. Wheatstone Bridge

The bridge is balanced when

$$R_1 R_4 = R_2 R_3$$

AC bridge is balanced when

$$Z_1 Z_4 = Z_2 Z_3$$

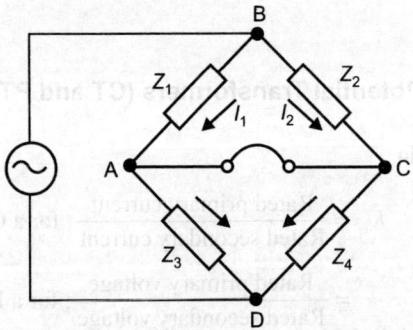

Fig. 15.1

7. Maxwell Bridge, $R_x = \dfrac{R_2 R_3}{R_1}$

$$L_x = R_2 R_3 C_1$$

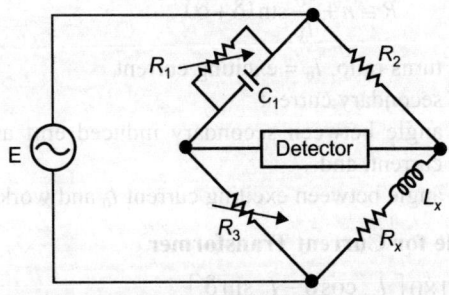

Fig. 15.2

8. Hay Bridge

$$R_x = \frac{\omega^2 C_1^2 R_1 R_2 R_3}{1 + \omega^2 C_1^2 R_1^2}$$

$$L_x = \frac{R_2 R_3 C_1}{1 + \omega^2 C_1^2 R_1^2}$$

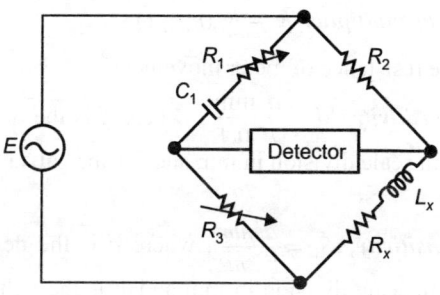

Fig. 15.3

9. Schering Bridge

The value of R_x and C_x are given by

$$R_x = \frac{R_2 C_1}{C_3}, \ C_x = C_3 \frac{R_1}{R_2}$$

$$\text{Power factor} \cong \frac{R_x}{X_x} \cong \omega C_x R_x$$

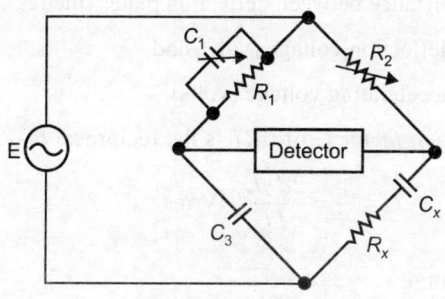

Fig. 15.4

10. Instruments: Shunts and multipliers are used to extend the range of ammeters and voltmeters.

Resistance of shunt $R_{sh} = \dfrac{R_m}{(m-1)}$

where *m* for shunt is given by:

$$m = \frac{\text{Total current}}{\text{Current for full scale deflection}}$$

Resistance of multiplier $R_s = R_m(m-1)$

where R_m is the resistance of meter movement

Current Sensitivity $SI = \dfrac{d}{I}\dfrac{\text{mm}}{\mu A}$, where *d* is the deflection of the galvanometer in scale division in mm and *I* is the galvanometer current in μA.

Voltage Sensitivity, $S_v = \dfrac{d}{V}\dfrac{nm}{m}$, where *d* is the deflection of the galvanometer in scale division in mm and *V* is the voltage applied to galvanometer in mV.

11. Deflection Sensitivity of CRT: It is defined as deflection on the screen (in metre) per volt of deflection voltage and is given by

$$S = \frac{D}{E_d} = \frac{Ll_d}{2dE_a}\,(\text{m/V})$$

where   *D* = deflection on the fluorescent screen (metres),

   L = distance from centre of deflection plates to screen (metres),

   l_d = effective length of deflection plates (metres),

   d = distance between deflection plates (metres),

   E_d = deflection voltage (volt), and

   E_a = accelerating voltage (volts).

The *deflection factor G* of *CRT* is the reciprocal of sensitivity.

$$G = \frac{1}{S} = \frac{2d \cdot E_a}{Ll_d}\,(\text{V/m})$$

12. Strain Gauge

Gauge Factor: The *gauge factor* is defined as unit change in resistance to unit change in length.

$$GF = \frac{dR/R}{dL/L} = \frac{dR/R}{\varepsilon_a}$$

$$= 1 + 2\mu + \frac{d\rho/\rho}{dL/L}$$

Reynolds Number $R = \dfrac{V \cdot D \cdot \rho}{\mu}$

where V = velocity, D = inside diameter of pipe,

  ρ = fluid density, and μ = viscosity.

13. Rotometer

Rate of flow $Q = KA_m 2g\sqrt{\dfrac{V_f}{A_f}\left(\dfrac{\rho_t}{\rho} - 1\right)}$

where,

  Q = rate of flow

  K = taper

  A_m = area between float and tube measured at bottom of float

  V_f = volume of float

  A_f = area of flat

  ρ_t = density of float, and ρ is the density of flowing fluid

Chapter 16

Control Engineering

1. Terminology of the Closed-loop Block Diagram

The terms used in the closed-loop block diagram shown in Fig. 16.1 are as follows:

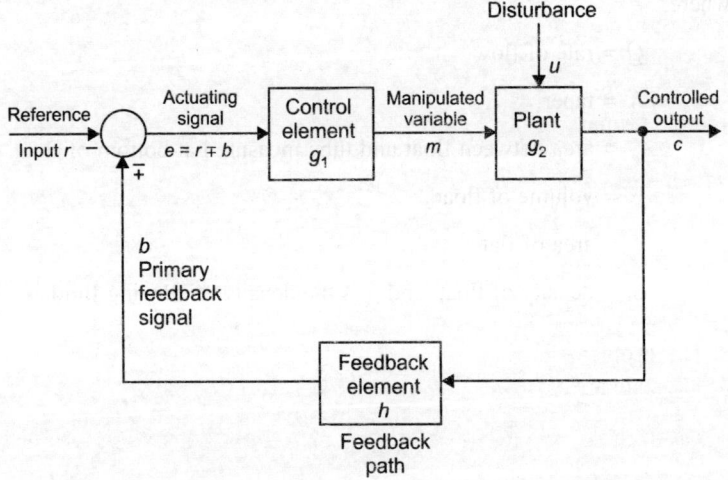

Fig. 16.1

The plant g_2, also called the *controlled system* is the body, process or machine of which a particular quantity or condition is to be controlled.

The control elements g_1, also called the *controller*, are the components required to generate the appropriate control signal m applied to the plant.

The *feedback elements h*, are the components required to establish the functional relationship between the primary feedback signal *b* and the controlled output *c*.

The *reference input r*, is an external signal applied to a feedback control system in order to command a specified action of the plant.

The *controlled output c*, is that quantity or condition of the plant which is to be controlled.

The *primary feedback signal b* is a signal which is a function of the controlled output *c*, and which is algebraically summed with the reference input *r* to obtain the actuating signal *e*.

The *actuating signal e*, also called the *error or control action*, is the algebraic sum of the reference input *r* and (plus or minus) the primary feedback *b*.

The *manipulated variable m*, (control signal) is that quantity which the control elements g_1 apply to the plant g_2.

A *disturbance u*, is an undesired input signal which affects the value of the controlled ouput *c*.

The *forward path* is the transmission path from actuating signal *e* to the controlled output *c*.

The *feedback path* is the transmission path from the controlled output *c* to the primary feedback signal *b*.

2. Automatic Control

Closed loop automatic control is operated by the comparison of the quantity to be controlled with the desired or reference value. The difference between the measured and desired values is used to actuate the controller (Fig. 16.2).

The problem in design is to obtain sufficient accuracy under both steady-state and transient conditions, with adequate speed of response.

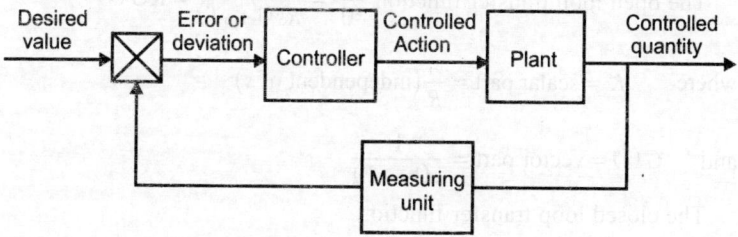

Fig. 16.2

This must be achieved without producing instability. Calculations fall into two parts: (a) analysis of existing "known" parts of the system; (b) synthesis of unknown part to give required performance. These are carried usually by frequency response methods, using transfer functions.

3. Transfer Function

It is the ratio of output to input expressed in terms of Laplace transform.

$$G(s) = \frac{C(s)}{R(s)} = \frac{b_m s^m + b_{m-1} s^{m-1} + \ldots + b_0}{a_n s^n + a_{n-1} s^{n-1} + \ldots + a_0}$$

Take angular position control as an example shown in Fig. 16.3

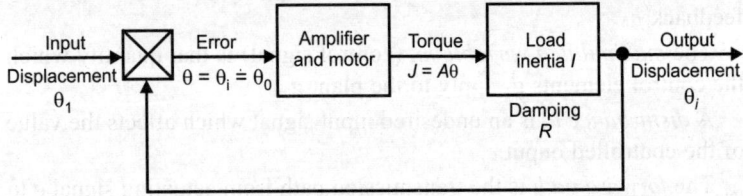

Fig. 16.3

For a viscous damped inertia load

$$\text{torque } J = I \frac{d^2\theta_0}{dt^2} + \frac{R\,d\theta_0}{dt}, \text{ using } s = d/dt$$

$$= (Is^2 + Rs)\theta_0$$

The transfer function, $\theta_0/J = \dfrac{1}{Rs(1+sT)}$

where T is time constant of the load $= I/R$

The open loop transfer function $\dfrac{\theta_0}{\theta} = \dfrac{A}{Rs(1+sT)} = KG(s)$

where $K = $ scalar part $= \dfrac{A}{R}$ (independent of s)

and $G(s) = $ vector part $= \dfrac{1}{s(1+sT)}$

The closed loop transfer function

$$\frac{\theta_0}{\theta_i} = \frac{KG}{1+KG}$$

Here $\quad \dfrac{\theta_0}{\theta_i} = \dfrac{A}{Rs(1+sT)+A} = \dfrac{A}{s^2 I + Rs + A} = \dfrac{\omega_n^2}{s^2 + 2\zeta\omega_n s + \omega_n^2}$

where $\quad \omega_n^2 = \dfrac{A}{I}, \; \zeta\omega_n = \dfrac{R}{2I}$

For sinusoidal signals, represented by rotating vectors $\theta_i e^{j\omega ts}$, $\theta_0 e^{(j\omega t\theta)}$ etc. $s = d/dt = j\omega$ and the transfer functions are:

$$\frac{\theta_0}{\theta} = \left|\frac{\theta_0}{\theta}\right| \angle \frac{\theta_0}{\theta} = \frac{A}{Rj\omega(1+j\omega T)}$$

$$\frac{\theta_0}{\theta_i} = \left|\frac{\theta_0}{\theta_i}\right| \angle \frac{\theta_0}{\theta_i} = \frac{\omega^2}{(j\omega)^2 + 2\xi\omega_n j\omega + \omega_n^2}$$

$$\left|\frac{\theta_0}{\theta_i}\right| = \frac{A}{R\omega\sqrt{1+\omega^2 T^2}} \angle \frac{\theta_0}{\theta} = -(90° + \tan^{-1}\omega T)$$

$$\left|\frac{\theta_0}{\theta_i}\right| = \frac{\omega_n^2}{\sqrt{\left(\omega_n^2 - \omega^2\right)^2 + 4\zeta^2\omega_n^2 \cdot \omega^2}} = \frac{1}{\sqrt{\left(1-f^2\right)^2 + 4\zeta^2 f^2}}$$

where f = frequency ratio = $\dfrac{\omega}{\omega_n}$ and $\angle \dfrac{\theta_0}{\theta_i} = -\tan^{-1}\dfrac{2\zeta}{1-f^2}$

These give the frequency response curve of the system with open and closed loops.

Vector Locus or Nyquist Diagram

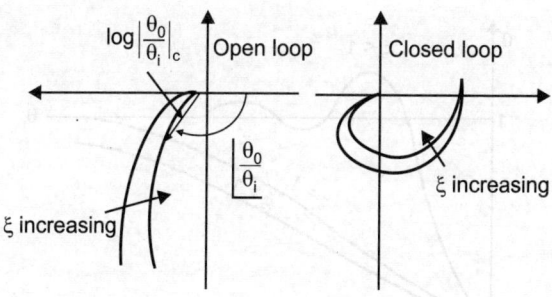

Fig. 16.4

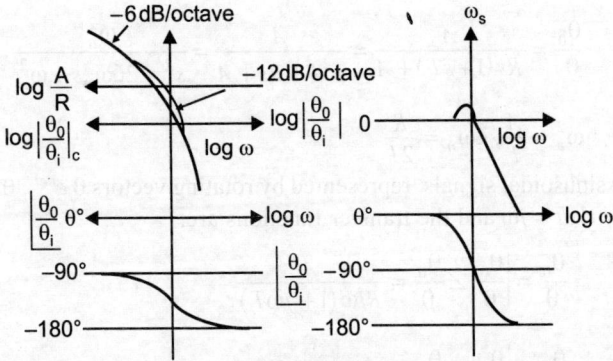

Fig. 16.5

The transfer function are plotted as the locus of the end of the vector, for ω between zero and infinity. For the system just analyzed, are as shown in Figs 16.4 and 16.5.

Transient Response

If input of the above system is given a unit step $\theta_i = 1(t)$, the response of the output for $\zeta < 1$ is shown in Fig. 16.6.

$$\theta_0 = 1(t) - \frac{\omega_n}{m} - e^{-\zeta\omega_n t} \cdot \cos(mt + \phi)$$

where $m^2 = \omega_n^2(1 - \zeta^2)$ and $\phi = \tan^{-1}\left(\frac{\zeta\omega_n}{m}\right)$

The first overshoot is $100 \cdot e^{\frac{-\zeta\omega_n}{m}t}$ per cent of the input step, and the time to each of the first peak is π/m.

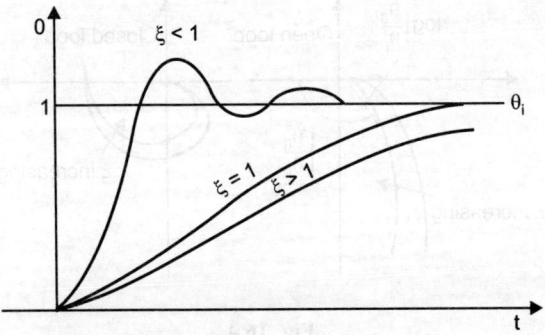

Fig. 16.6

4. Mason's Gain Formula for Singal Flow Graph

The transmittance between an input node and output node is the overall gain or overall transmittance. Between these two nodes, Mason's gain formula, which is applicable to the overall gain is given by

$$P = \frac{1}{\Delta} \sum_k P_k \Delta_k$$

where,

P_k = path gain transmittance of *the kth forward path*

Δ = Determinant of the graph

= 1 – (sum of all different loop gains)

+ (sum of gain products of all possible combinations of two non-touching loops)

– (sum of gain products of all possible combinations of three non-touching loops)

$$= 1 - \sum_a L_a + \sum_{b,c} L_b L_c - \sum_{d,e,f} L_d L_e L_f$$

Δ_k = Cofactor of the kth path determinant of this graph with the loops touching the kth forward path removed.

5.0 Analogous Quantities

Table 5.1   Force–Voltage Analogy

Mechanical System	Electrical System
Force	Voltage
Mass	Inductance
Viscous friction coefficient	Resistance
Spring constant	Reciprocal of capacitance
Displacement	Charge
Velocity	Current

Table 5.2   Force–Current Analogy

Mechanical System	Electrical System
Force	Current
Mass	Capacitance
Viscous–friction coefficient	Reciprocal of resistance
Spring constant	Reciprocal of inductance
Displacement	Magnetic flux linkage
Velocity	Voltage

5.1 Sensors and Encoders in Control System

Sensors and encoders are important components used to monitor the performance and for feedback control systems. The sensors and encoders that are commonly used in control systems are:

(i)   Potentiometer

(ii) Synchros

(iii) Tachometers

(iv) Incremental encoder.

Synchros are widely used in control system as detector and encoders because of their ruggedness in construction and high reliability. Incremental encoders are frequently found in modern control systems.

6. Block Reduction

Any finite number of blocks in series may be algebraically combined by multiplication. That is n blocks with transfer functions G_1, G_2, ..., G_n connected in cascade are equivalent to single element G with transfer function given by

$$G = G_1 \cdot G_2 \cdot G_3 \dots G_n$$

Various simplifications in the block diagrams are illustrated in Fig. 16.7.

Block diagram **Equivalent block diagram**

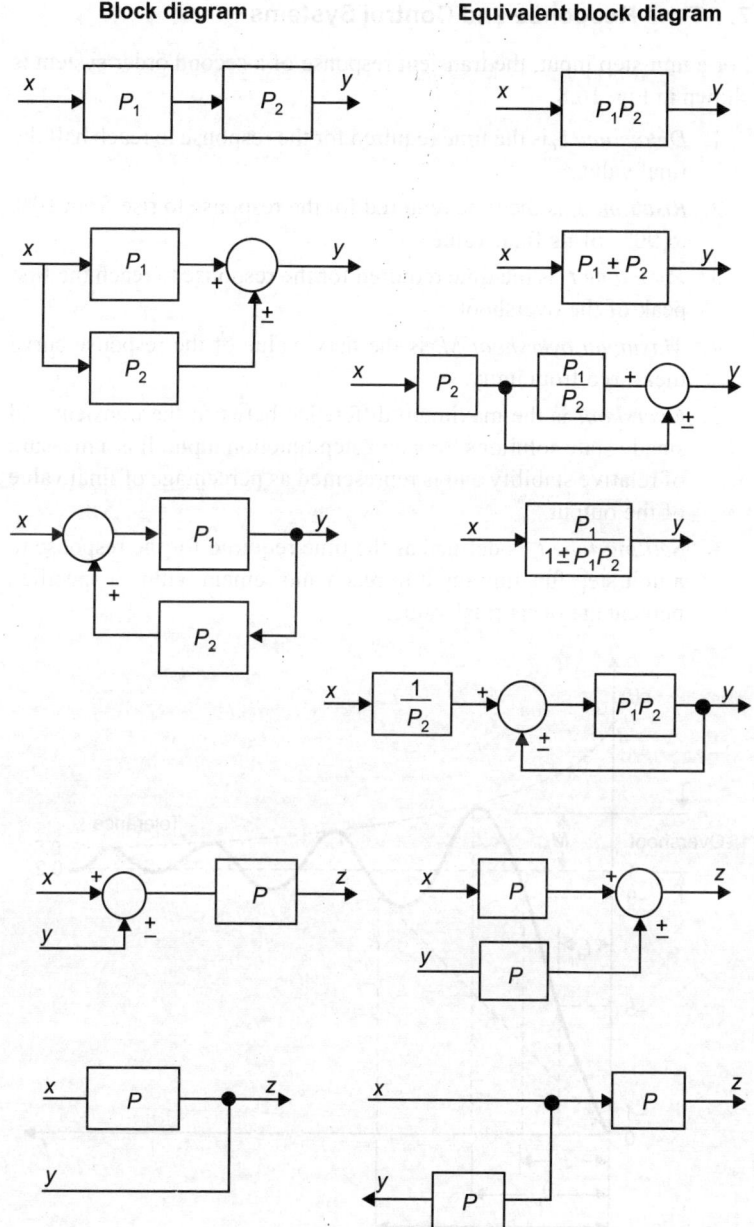

Fig. 16.7

7. Time Response and Control Systems

For a unit step input, the transient response of a second order system is shown in Fig. 16.8.

1. *Delay time t_d* is the time required for the response to reach half the final value.

2. *Rise time t_r* is the time required for the response to rise from 10% to 90% of its final value.

3. *Peak time t_p* is the time required for the response to reach the first peak of the overshoot.

4. *Maximum overshoot M_p* is the max. value of the response curve measured from unity.

5. *Overshoot* is the maximum difference between the transient and steady state solutions for a unit step function input. It is a measure of relative stability and is represented as percentage of final value of the output.

6. *Settling time t_s* is defined as the time required for the response to a unit step function input to reach and remain within a specified percentage of its final value.

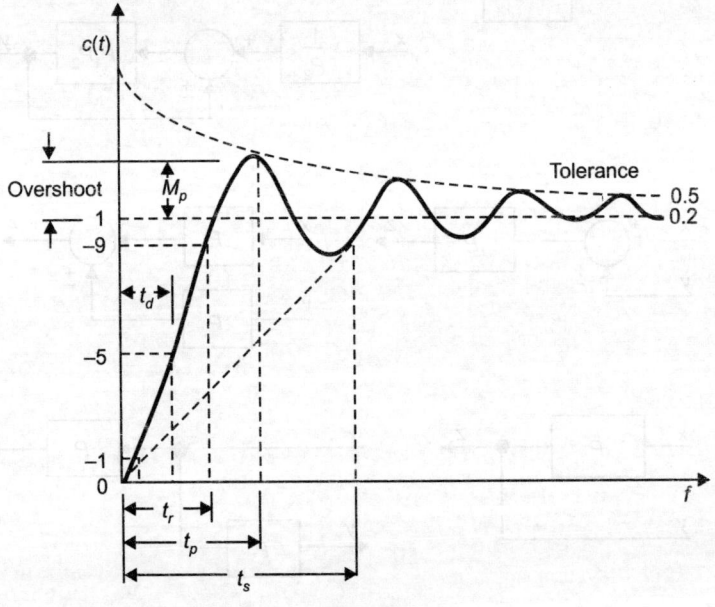

Fig. 16.8

8. Test Functions

Typical test signals: commonly used signals are step function, ramp function, impulse function and sinusoidal function.

A unit step function shown in Fig. 16.9a is a function of time denoted by $u(t - t_0)$ and defined by

$$\mu(t - t_0) = \begin{cases} 1 & t > t_0 \\ t & t \le t_0 \end{cases}$$

A unit ramp function shown in Fig. 16.9b is a function of time which is the integral of unit step function given by

$$\int_{-\infty}^{t} u(\tau - t_0)d\tau = \begin{cases} t - t_0 & t > t_0 \\ 0 & t \le t_0 \end{cases}$$

A unit impulse function $\delta(t)$ shown in Fig. 16.9c may be defined by

$$= \lim_{\substack{\Delta t \to 0 \\ \Delta t > 0}} \left[\frac{u(t) - u(t - \Delta t)}{\Delta t} \right]$$

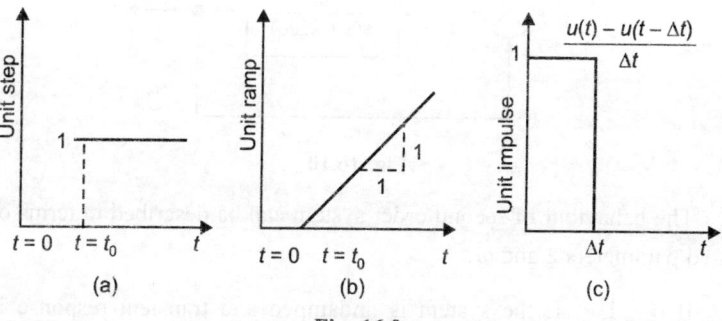

Fig. 16.9

where $u(t)$ is the unit step function.

Area under the curve = 1 for all $\Delta t \int_{-\alpha}^{\alpha} \delta(t)dt = 1$

9. Order of Control

This is number of integrations or 1/s factors in the open loop transfer function KG. This affects steady state errors.

Order 0 system has a displacement error $\dfrac{1}{1+K}$ for constant input displacement.

Order 1 system has zero displacement error but has a velocity error.

proportional to $\dfrac{1}{K}d\theta_i/dt$ for constant input velocity $d\theta_i/dt$.

Order 2 has zero velocity error, but acceleration error proportional to

$\dfrac{1}{K}\dfrac{d^2\theta_i}{dt^2}$.

All can be reduced by increasing sensitivity K.

Second Order System

The second order system is shown in Fig. 16.10 which has transfer function

$$\frac{C(s)}{R(s)}=\frac{\omega_n^2}{s^2+2\xi\omega_n+\omega_n^2}$$

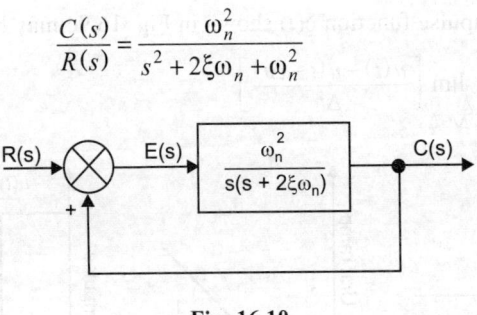

Fig. 16.10

The behaviour of second order system can be described in terms of two parameters ξ and ω_n.

If $0<\zeta<1$, the system is undamped and transient response is oscillatory. If $\zeta=1$ the system is called *critically damped*. An over-damped system corresponds to $\xi>1$.

Step response of second-order system is

$$c(t)=1+\frac{1}{2\sqrt{\xi^2-1}\left(\xi+\sqrt{\xi^2-1}\right)}e^{-\left(\xi+\sqrt{\zeta^2-1}\right)\omega_n t}$$

Error Constants. The error constants and steady errors for type 0, type 1 and type 2 unity feedback systems are tabulated as follows.

Input	Unit step		Unit ramp		Unit parabola	
System type	K_p	Steady state error	K_v	Steady state error	K_a	Steady state error
Type 0	$\dfrac{K\beta_1(0)}{\beta_2(0)}$	$\dfrac{1}{1+K_P}$	0	∞	0	∞
Type 1	∞	0	$\dfrac{K\beta_1(0)}{\beta_2(0)}$	$\dfrac{1}{K_V}$	0	∞
Type 2	∞	0	∞	0	$\dfrac{K\beta_1(0)}{\beta_2(0)}$	$\dfrac{1}{K_a}$

10. Complex Plane: Pole-Zero Maps

The rational function $F(s)$ can be written as

$$F(s) = \frac{b_m \prod\limits_{t=1}^{m}(s+z_i)}{\prod\limits_{i=1}^{n}(s+p_i)}$$

where the terms $(s+z_i)$ are factors of the numerator polynomial and the terms $(s+p_i)$ are the factors of denominator polynomial.

Zeros: Those values of the complex variables for which $|F(s)|$ is zero are called *zeros of F(s)*.

Poles: Those values of the complex variables for which $|F(s)|$ is infinite are called *poles of F(s)*.

11. Stability

The stability of a system is determined by its response to input or disturbances. A stable system is one that will remain at rest unless excited by an external source and will return to rest *if* all the excitations are removed.

Marginally Stable: The system has some roots with real parts equal to zero, but none with positive real parts, the system is said to be *marginally stable*, which is unstable.

12. Routh Stability Criterion

It is a method for determining system stability that can be applied to an nth order characteristic equation of the form

$$a_n s^n + a_{n-1} s^{n-1} + \ldots + a_1 s + a_0 = 0$$

The criterion is applied through the use of Routh table defined as follows:

s^n	a_n	a_{n-2}	$a_{n-4}\ldots\ldots$
s^{n-1}	a_{n-1}	a_{n-3}	$a_{n-5}\ldots\ldots$
	b_1	b_2	b_3
	c_1	c_2	c_3
	$\ldots\ldots$	$\ldots\ldots$	$\ldots\ldots$

where a_n, a_{n-1}, a_0 are the coefficients of the characteristic equation and

$$b_1 = \frac{a_{n-1}a_{n-2} + a_n a_{n-3}}{a_{n-1}}; \quad b_2 = \frac{a_{n-1}a_{n-4} - a_n a_{n-5}}{a_{n-1}}$$

$$c_1 = \frac{b_1 a_{n-3} - a_{n-1} b_2}{b_1}; \quad c_2 = \frac{b_1 a_{n-5} - a_{n-1} b_3}{b_1}$$

All the roots of the characteristic equation have negative real parts if and only if the elements of the first column of the Routh table have the same sign.

13. The Nyquist Stability Criterion

A closed-loop control system is absolutely stable if the roots of the characteristic equation have negative real parts. Equivalently, the poles of the closed-loop transfer function or the zeros of the denominator.

$$1 - GH(s)$$

of the closed-loop transfer function must lie in LHS of s-plane.

(i) *Relative Stability*: The relative stability of a feedback control system is determined from the polar or Nyquist stability plot.

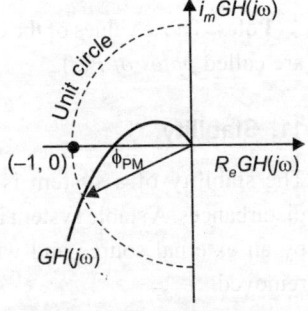

Fig. 16.11

The *phase crossover frequency* ω_π is that frequency at which the phase angle of $GH(j\omega)$ is $-180°$, that is, the frequency at which the polar plot crosses the negative real axis, the gain margin

$$= \frac{1}{|GH(j\omega_\pi)|}$$

Gain crossover frequency ω is that frequency at which $|GH = (j\omega)| = 1$.

Phase Margin (ϕ_{PM}) is the angle by which polar plot must be rotated to cause it pass through $(-1, 0)$ point. It is given by

$$\phi_{PM} = [180° + arg. (GH\,j\omega)] \text{ degrees}$$

(ii) **Stability:** Consider system with open loop transfer function

$\dfrac{K}{s(1+sT_1)(1+sT_1)}$ whose Nyquist diagram is shown in Fig. 16.12.

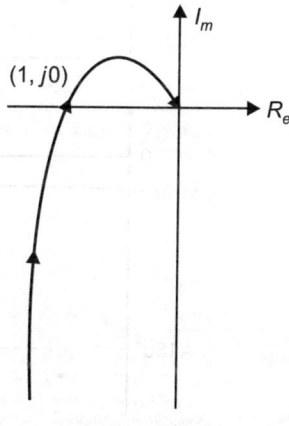

This condition is harmonic instability. If K is increased, oscillations will build up and the system is stable. For stability, the point $(-1, j0)$ must lie to the left of the locus, looking in the direction of increasing frequency. Generally, increasing K to reduce errors will tend to produce instability. A value of K which gives $\left|\dfrac{\theta_0}{\theta_i}\right| \max = 1.2$ to 1.3

Fig. 16.12

will give satisfactory stability. If K is adjusted so that the locus passes through the point $(-1, j0)$ at some frequency ω_1, then at this frequency $\theta_0/\theta = -1$ or $\theta = -\theta_0$. Since at any frequency $\theta = \theta_i - \theta_0$, a sinusoidal signal at frequency ω_1 can travel round the closed loop even with zero input θ_i.

14. Bode's Plot

A sinusoidal function may be represented by two separate plots. One giving the magnitude *vs* frequency and the other the phase angle *vs* frequency. A Bode's plot of logarithm of the magnitude of sinusoidal transfer function, and the other is a plot of the phase angle, both are plotted against the frequency in logarithmic scale.

The main advantage of Bode's plot is that multiplication of magnitudes can be converted into addition. It is the simplest method. Transfer function can be determined easily by Bode's plot, the low frequency and high frequency characteristics of the transfer function can be determined.

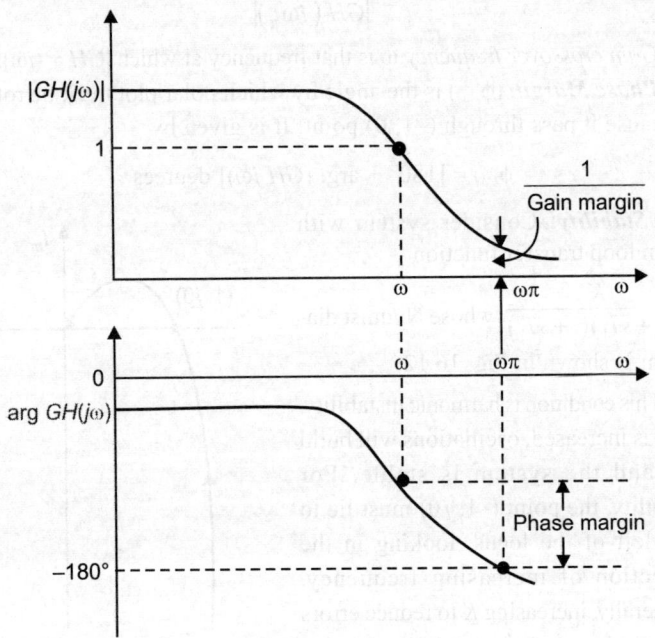

Fig. 16.13

15. Root Locus

The closed-loop transfer function of this system is

$$\frac{C}{R} = \frac{G}{1+GH}$$

(i) Centre of asymptotes is given by

$$\sigma_c = \frac{\sum\limits_{i=1}^{n} p_i - \sum\limits_{i=1}^{m} z_i}{n-m}$$

where p_i are the poles, $-z_i$ are the zeros, n is the number of poles, and m the number of zeros of GH.

(ii) Angle between the asymptotes and real axis are given by

$$\beta = \frac{(2l+1)180}{n-m} \text{ degree for } k > 0$$

$$\frac{2l \times 180}{n-m} \text{ degree for } k < 0 \text{ for } l = 0, 1, 2, 3, n - m + 1$$

(iii) Break away point can be determined by solving the equation

$$\sum_{i=1}^{n} \frac{1}{\sigma_b + p_i} = \sum_{i=1}^{m} \frac{1}{\sigma_b + z_i},$$

where $-p_i$ and $-z_i$ are the poles and zeros of GH respectively.

16. State Space Analysis of Control Systems

The state space approach is best suited for a complex system having many inputs and outputs. This technique reduces the complexity of mathematical expressions and includes the initial conditions in the design. The terms used in this technique are defined as below.

State: The state of a dynamic system is the smallest set of variables called *state variables* such that the knowledge of these variables at $t = t_0$, together with the input for $t > t_0$, completely determines the behaviour of the system for any time $t > t_0$.

State Variables: The state variables of a dynamic system are the smallest set of variables which determine the state of the dynamic system. If n variables $x_1(t)$, $x_2(t)$,..., $x_n(t)$ describe the behaviour of a dynamic system, then these n variables are a set of state variables.

State Vector: If n state variables are needed to completely describe the behaviour of a given system, then these n state variables can be considered to be the n components of a vector $x(t)$ such a vector is called a *state vector*.

State Space: The n-dimensional space whose coordinate axes consist of the X_1 axis, X_2 axis and X_n axis is called *state space*. Any state can be represented by a point in the state space.

State Space Representation of nth Order Differential Equation: Consider the equation

$$y^n + a_1 \overset{(n-1)}{y} + ... + ... + a_{n-1} \dot{y} + a_n y = u$$

where $\qquad y(t), \dot{y}(t), \overset{(n-1)}{y}(t)$ is a set of n state variables.

The above equation can be written as

$$\dot{x} = Ax + Bu$$

where $\quad x = \begin{bmatrix} x_1 \\ x_2 \\ . \\ . \\ . \\ x_n \end{bmatrix} \quad A = \begin{bmatrix} 0 & 1 & 0 & ... & 0 \\ 0 & 0 & 1 & ... & 0 \\ ... & ... & ... & ... & ... \\ 0 & 0 & 0 & ... & 1 \\ -a_n & -a_{n-1} & & & -a_1 \end{bmatrix} \quad B = \begin{bmatrix} 0 \\ 0 \\ 0 \\ 0 \\ 0 \\ 0 \end{bmatrix}$$

The output equation becomes

$$y = \begin{bmatrix} 1 & 0 & 0 & ... & 0 \end{bmatrix} \begin{bmatrix} x_1 \\ x_2 \\ . \\ . \\ . \\ x_n \end{bmatrix}$$

or $\qquad y = Cx$

where $\quad C = \begin{bmatrix} 1 & 0 ... 0 \end{bmatrix}$

Eigen values of $n \times n$ matrix A: The eigen values of an $n \times n$ matrix A are the roots of the characteristic equation $[\lambda I - A] = 0$
Eigen values are known as *characteristics roots*.

17. Improvement of Response by Cascade Elements

(a) Integral compensation by the element giving an output

$$\theta^l = K_1 \theta + K_2 \int \theta \, dt$$

$$= K_1 \theta \left(1 + \frac{1}{sT}\right) \text{ where } T = \frac{K_1}{K_2}$$

i.e. transfer function $\dfrac{\theta^l}{\theta} = \dfrac{K_1 (l + sT)}{sT}$

This is equivalent to an additional integration with a phase advance element $(1 + sT)T$ can be chosen to avoid producing instability.

(b) *Approximate integral compensation*

This network gives

$$\frac{V_0}{V_i} = \frac{1+sT}{1+asT}$$

where $a = \dfrac{R_1 + R_2}{R_2}$, $T = R_2C$.

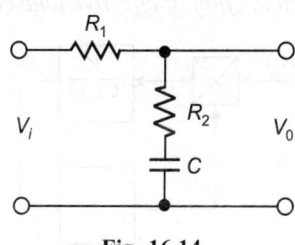

with $a = 10$, and a suitable value of T, the overall gain K can be increased by a factor of about 6, thus reducing the steady-state error in that ratio.

Fig. 16.14

Both forms of integral compensation slow down the response curve.

(c) *Phase advance network*

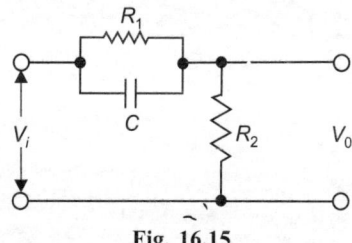

Fig. 16.15

$$\frac{V_0}{V_i} = \frac{1}{a}\frac{(1+asT)}{(1+sT)}$$

where $T = \dfrac{R_1 R_2 C}{R_1 + R_2}$

Max. phase advance $= \sin^{-1}\dfrac{a-1}{a+1}$

at $\omega = \dfrac{l}{T\sqrt{a}}$ and $a = \dfrac{R_1 + R_2}{R_2}$

This allows gain increase by a factor slightly less than 'a' and a corresponding reduction of steady-state error. It also speeds up the transient response by the same factor.

18. Improvement of Response by Feedback Elements

Reduction of effective time constant of a lagging element.

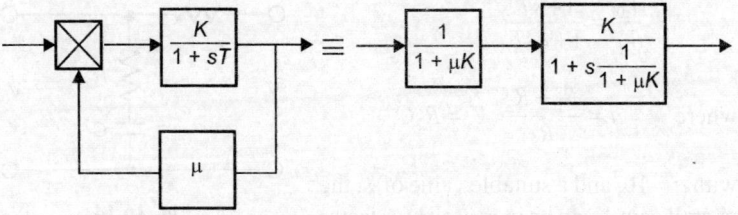

Fig. 16.16

Time constant and overall gain are reduced in the ratio $1/(1 + \mu K)$.
The loss of gain can be made-up elsewhere.

Chapter **17**

Generation, Transmission and Distribution

1. Characteristics of Turbines

Specific speed is defined as the speed of a geometrically similar turbine working under unit head and delivering unit power. The specific speed in British unit N_s is given by the formula

$$\frac{N\sqrt{HP}}{H^{5/4}} \text{ rpm}$$

where N = Rotational speed in rpm; HP = Horse power of the wheel and; H = Head of water in metre.

Further if power is expressed in kW, the above expression becomes

$$N_s = \frac{1.165\,N\sqrt{kW}}{H^{5/4}} \text{ rpm}$$

2. Calculation of HP (Metric) and kW Power

Let Q = Discharge in cu·m/sec

       H = Head of water; η = overall efficiency of hydrostation

       Since 1 m^3 of water weighs = 1000 kg · m

       Work done/sec   = 1000 QH kg · m/sec

       1 HP (metric)   = 75 kg · m/sec.

       HP developed $= \dfrac{1000\,Q \cdot H \cdot \eta}{75}$

$$\text{Power (kW)} = \frac{1000Q \cdot H \cdot \eta \times 735.5}{75 \times 1000}$$

$$P = 9.8 \, QH\eta$$

3. Units of Energy and Relationships

(i) Mechanical energy (Joules) = Force (newton) × distance (m)

(ii) Electrical energy (Joules) = Voltage (V) × Current (A) × time (sec)

(iii) 1 kilowatt hour (kWh) = 1 kW × 1 hour

1 kWh = 1000 (W) × 3600 (sec) = 36 × 10^5 watt-sec or Joules

1 kilo calorie = 1000 calories; 1 calorie = 4.18 Joules

4. Terms Commonly Used in System Operation

Firm Power: It is the power intended to be always available.

Cold Reserve: It is that reserve generating capacity which is available for service but is not in operation.

Hot Reserve: It is that reserve generating capacity which is in operation but is not in service.

Spinning Reserve: It is that generating capacity which is connected to bus and is ready to take load.

Connected Load: It is the sum of continuous ratings of all the equipment connected to the supply system.

Maximum Demand: It is the greatest demand of load on power station during a given period.

Demand Factor: It is the ratio of maximum demand on the power station to its connected load, *i.e.*

$$\text{Demand factor} = \frac{\text{Maximum demand}}{\text{Connected load}}$$

Average Load: The average of loads occurring on the power station in a given period is known as average load or average demand.

$$\text{Daily average load} = \frac{\text{No. of units (kWh) generated in a day}}{24 \text{ hours}}$$

$$\text{Monthly average load} = \frac{\text{No. of units generated in a month}}{\text{No. of hours in a month}}$$

$$\text{Yearly average load} = \frac{\text{No. of units generated in a year}}{8760 \text{ hours}}$$

Load Factor: The ratio of average load to the maximum demand during a given period is known as *load factor*.

$$\text{Load factor} = \frac{\text{Average load}}{\text{Maximum demand}}$$

Diversity Factor: The ratio of the sum of individual maximum demands to the maximum demand on the power station is known as *diversity factor*.

$$\text{Diversity factor} = \frac{\text{Sum of individual maximum demands}}{\text{Maximum demand on power station}}$$

$$\text{Capacity factor} = \frac{\text{Actual energy produced}}{\text{Maximum energy that could have been produced}}$$

$$= \frac{\text{Average demand}}{\text{Plant capacity}}$$

Reserve capacity = Plant capacity – maximum demand

Utilization Factor (Plant Use Factor): It is the ratio of kWh generated to the product of plant capacity and the number of hours for which the plant was in operation.

$$\text{Plant use factor} = \frac{\text{Station output in kWh}}{\text{Plant capacity} \times \text{Hours of use}}$$

Table 17.1   Comparison of Various Systems of Transmission

System	Same Maximum Voltage to Earth	Same Maximum Voltage between Conductors
(a) DC System		
(i) Two-wire	1	1
(ii) Two-wire midpoint earthed	0.25	1
(iii) Three-wire	0.3125	1.25
(b) Single Phase System		
(i) Two-wire	$\dfrac{2}{\cos^2 \phi}$	$\dfrac{2}{\cos^2 \phi}$

(Contd...)

Table 17.1 Comparison of Various Systems of Transmission (*Contd.*)

System	Same maximum voltage to earth	Same maximum voltage between conductors
(ii) Two-wire with mid-point earthed	$\dfrac{0.5}{\cos^2\phi}$	$\dfrac{2}{\cos^2\phi}$
(iii) Three-wire	$\dfrac{0.625}{\cos^2\phi}$	$\dfrac{2.5}{\cos^2\phi}$
(c) Two-Phase System		
(i) Two-Phase, four-wire	$\dfrac{0.5}{\cos^2\phi}$	$\dfrac{2}{\cos^2\phi}$
(ii) Two-Phase, three-wire	$\dfrac{1.457}{\cos^2\phi}$	$\dfrac{2.194}{\cos^2\phi}$
(d) Three-Phase System		
(i) Three-phase, three-wire	$\dfrac{0.5}{\cos^2\phi}$	$\dfrac{1.5}{\cos^2\phi}$
(ii) Three-phase, four-wire	$\dfrac{0.583}{\cos^2\phi}$	$\dfrac{1.75}{\cos^2\phi}$

5.   Kelvin's Law: Economic Choice of Conductor Size

The most economical area of conductor is that for which the total annual cost of transmission line is minimum. This is known as *Kelvin's Law*. The total annual cost of transmission line can be divided broadly into two parts, *i.e.* annual charge on capital outlay and annual cost of energy wasted in the conductor.

Total annual cost = $(P_1 + P_2\, a) + P_3 \neq a$

where P_1, P_2 and P_3 are constants and 'a' is the area of cross-section of the conductor.

6.   The Empirical Formula for Economic Voltage for Transmission Line

$$V = 5.5\sqrt{0.62l + \frac{3P}{150}}$$

where V is line voltage in kV, l is distance of transmission line in km, P is the maximum power in kW/phase to be delivered to a single circuit.

7. Parameters

Resistance: The DC resistance R of a conductor of length l and cross-sectional area A is

$$R = \rho \frac{l}{a} \text{ ohm}$$

where ρ is the *resistivity* of the material of the conductor in ohm-metres. The temperature dependence of resistance is quantified by the relation

$$R_2 = R_1[1 + \alpha(T_2 - T_1)]$$

where R_1 and R_2 are the resistance at temperature T_1 and T_2, respectively, and α is called the *temperature coefficient of resistance*. The resistivities and temperature coefficients of several metals are given in Table 17.2.

Long transmission lines may involve shunt resistance (or conductance) in addition to series resistances.

Table 17.2   Resistance and Temperature Coefficients of Resistance

Material	Resistivity ρ at 20°C $\mu\omega \cdot cm$	Temperature Coefficient α at 20°C
Aluminium	2.83	0.0039
Brass	6.4–8.4	0.0020
Copper:		
Hard-drawn	1.77	0.00382
Annealed	1.72	0.00393
Iron	10.0	0.0050
Silver	1.59	0.0038
Steel	12–88	0.001–0.005

8. Inductance of Two-wire, Single-phase Line

The inductance per conductor of a two-wire, single-phase transmission line is given by

$$L_1 = \frac{\mu_0}{8\pi}\left(1 + 4\ln\frac{D}{r}\right) \text{ henry/metre}$$

where $\mu_0 = 4\pi \times 10^{-7}$ H/m (the permeability of free space). D is the distance between the centres of the conductors, and r is the radius of the conductor.

The loop, inductance is given by

$$L = 2L_1 = \frac{\mu_0}{4\pi}\left(1 + 4\ln\frac{D}{r}\right) = \left(1 + 4\ln\frac{D}{r}\right) \times 10^{-7} \text{ H/m}$$

Since $\ln e^{1/4} = {}^1/_4$, the above equation may also be written as

$$L = 4 \times 10^{-7} \ln\frac{D}{r'} \text{ H/m}$$

where $r' = re^{-1/4}$ is known as the geometric mean radius (GMR) of the conductor.

9. Inductance of Three-Wire, Three-Phase Line

The per-phase (or line-to-neutral) inductance of a three-phase transmission line with equilaterally space conductors is

$$L = \frac{\mu_0}{8\pi}\left(1 + 4\ln\frac{D}{r}\right) = 2\left(\frac{1}{4} + \ln\frac{D}{r}\right) \times 10^{-7} \text{ H/m}$$

where r is the conductor radius and D is the spacing between conductors.

10. Inductance of Composite Conductors

For a single-phase line consisting of two composite conductors, X and Y, the inductance L_x of conductor X is

$$L_x = 2 \times 10^{-7} \ln\frac{\sqrt[mn]{(D_{aa'}D_{ab'}D_{ac'}\dots D_{am})(D_{ba'}D_{bb'}D_{bc'}\dots D_{bm})\dots(D_{ma'}D_{nb'}D_{nc'}\dots D_{nm})}}{\sqrt[n^2]{(D_{aa}D_{ab}D_{ac}\dots D_{an})(D_{ba}D_{bb}D_{bc}\dots D_{bn})\dots(D_{na}D_{nb}D_{nc}\dots D_{nm})}} \text{ (H/m)}$$

where $D_{kk} = r'_k = r_k e^{-1/4}$ is the geometric mean radius (GMR) of the kth conductor.

The inductance L_Y of conductor Y is similar as L_X. The total line inductance becomes

$$L_X = L_x + L_Y$$

11. Capacitance

The shunt capacitance per unit length of a single-phase, two-wire transmission line is given by

$$C = \frac{\pi\varepsilon_0}{\ln(D/r)} \text{ farad/metre}$$

where ε_0 is the permittivity of free space. For a three-phase line with equilaterally space conductors, the per-phase (or line-to-neutral) capacitance is

$$C = \frac{2\pi\varepsilon_0}{\ln(D/r)} \text{ F/m}$$

The per-phase capacitance for the double-circuit transmission line is given by

$$C = \frac{4\pi\varepsilon_0}{\ln\left[\sqrt[3]{2}\,(D/r)(G/F)^{2/3}\right]}\ \text{F/m}$$

Using the concept of the GMD, we may write the capacitance to neutral of a nonsymmetrical three-phase double-circuit line as

$$C_n = \frac{2\pi\varepsilon_0}{\ln(\text{GMD/GMR})} = \frac{2\pi\varepsilon_0}{\ln(D_m/D_s)}\ \text{F/m}$$

$$C_n = \frac{10^{-9}}{18\ln(D_m/D_s)}\ \text{F/m}$$

12. Short-Transmission Line

The short transmission line is represented by the lumped parameters R and L, as shown in Fig. 17.1. Notice that R is the resistance per phase and L is the inductance per phase of the *entire line*. The line is shown to have two ends: the sending end designated by the subscript s at the generator, and the receiving end designated R at the load. Quantities of significance here are the voltage regulation and efficiency of transmission for lines of all lengths are defined as follows:

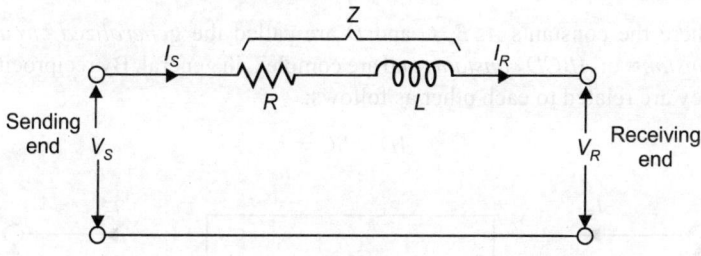

Fig. 17.1

$$\text{Voltage regulation} = \frac{|V_S - V_R|}{|V_R|} \times 100$$

Efficiency of transmission

$$\eta = \frac{\text{power at receiving end}}{\text{power at sending end}} = \frac{P_R}{P_S}$$

where V_R is the receiving-end voltage.

13. Long Transmission Line

The parameters of a long line are considered to be distributed over the entire length of line L

$$V_S = V_R \cosh\gamma L + I_R Z_C \sinh\gamma L$$

$$I_S = \frac{V_R \sinh\gamma L}{Z_C} + I_R \cosh\gamma L$$

$$\cosh\gamma L = 1 + \frac{(\gamma L)^2}{2!} + \frac{(\gamma L)^4}{4!} + \ldots + \cong 1 + \frac{1}{2}yz$$

$$\sinh\gamma L = 1 + \frac{(\gamma L)}{3!} + \frac{(\gamma L)^5}{5!} + \ldots + \cong \sqrt{yz}\left(1 + \frac{1}{6}yz\right)$$

where γ is known as the propagation constant and $\gamma = \sqrt{yz}$, y is the shunt admittance per unit length of line, z is the series impedance per unit length.

14. Transmission Line as a Two-port Network

$$V_S = AV_R + BI_R$$

$$I_S = CV_R + DI_R$$

where the constants A, B, C and D are called the *generalized circuit constants* or *ABCD constants* and are complex, in general. By reciprocity they are related to each other as follows:

$$AD - BC = 1$$

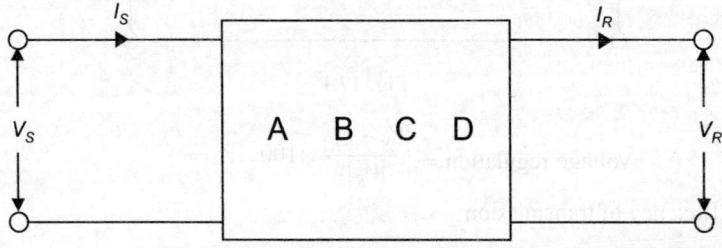

Fig. 17.2

15. ABCD Constants for Transmission Lines

Line Length	Equivalent Circuit	A	B	C	D
Short	Series impedance	1	Z	0	1
Medium	Nominal $-\pi$,	$1 + \frac{1}{2}YZ$	Z	$Y\left(1 + \frac{1}{4}YZ\right)$	$1 + \frac{1}{2}YZ$
	Nominal-T	$1 + \frac{1}{2}YZ$	$Z\left(1 + \frac{1}{4}YZ\right)$	Y	$1 + \frac{1}{2}YZ$
Long	Distributed parameter	$\cos\gamma L$	$Z_c \sinh\gamma L$	$(\sinh\gamma L)/Z_c$	$\cosh\gamma L$

Power Flow on Transmission Lines

The complex power $V_R I_R^*$ at the receiving end is thus given by

$$P_R + jQ_R = \frac{|V_R||V_S|}{|B|}|B|\angle\beta - \delta - \frac{|A|\,|V_R|^2}{|B|}\angle\beta - \alpha$$

So that

$$P_R = \frac{|V_R||V_S|}{|B|}\cos(\beta - \delta) - \frac{|A|\,|V_R|^2}{|B|}\cos(\beta - \alpha)$$

$$Q_R = \frac{|V_R||V_S|}{|B|}\sin(\beta - \delta) - \frac{|A||V_R|^2}{|B|}\sin(\beta - \alpha)$$

16. Traveling Waves on Transmission Lines

The ratio $\sqrt{L/C}$ has the dimension of ohm and is called the *characteristic impedance* Z_C of the line.

$$Z_C = \sqrt{L/C} = R_C \text{ for a lossless line}$$

If only forward-traveling waves exist at the load, then

$$V^+\left(t - \frac{L}{\mu}\right) = R_C I^+\left(t - \frac{L}{u}\right)$$

and if only backward-traveling waves exist at the load, then

$$V^-\left(t - \frac{L}{u}\right) = -R_C I^-\left(t - \frac{L}{u}\right)$$

17. Reflection Coefficients

The current reflection coefficients is the negative of the voltage reflection coefficients.

Source voltage reflection coefficient

$$\Gamma_S = \frac{R_S - R_C}{R_S + R_C}$$

18. Ferranti Effect

When the long transmission line having high capacitance is unloaded or is operated at light loads, it will be observed that the voltage at the receiving end is more than that of sending end. This phenomenon of rise of voltage at the receiving end of the lightly loaded or unloaded line is called as *Ferranti effect*. The rise in voltage is due to the fact that a charging current flowing in the line leads from the receiving end voltage by 90° (assuming that the capacitance of the line is concentrated at the receiving end).

19. Grading of Cables

The cable in Fig. 17.3(a) is filled with a single layer of a single dielectric, but many cables contain several layers of dielectrics. The dielectric materials in such a cable are chosen and distributed so as to minimize the difference between the maximum and minimum electric field strengths

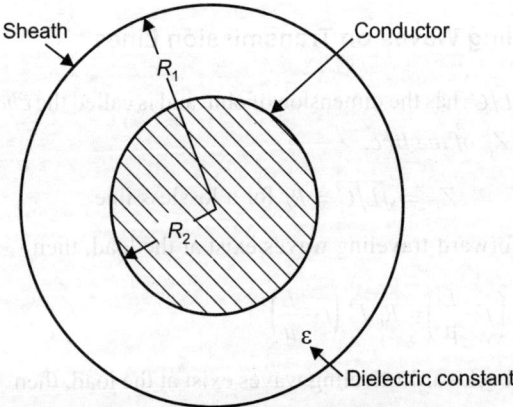

Fig. 17.3(a)

in the cable. This process is known as *grading*. Two types are commonly used in grading: capacitance grading and intersheath grading.

Capacitance Grading: In capacitance grading, two or more layers of different dielectric are used to insulate a cable, *e.g.* two such layers are shown in Fig. 17.3(b). The permittivities of these layers are so chosen that the maximum field strength is the same in both regions. The corresponding variation of the electric field E with radius r is shown in Fig. 17.3(b). For equal maximum field strengths,

$$\varepsilon_1 R_2 = \varepsilon_2 R_3$$

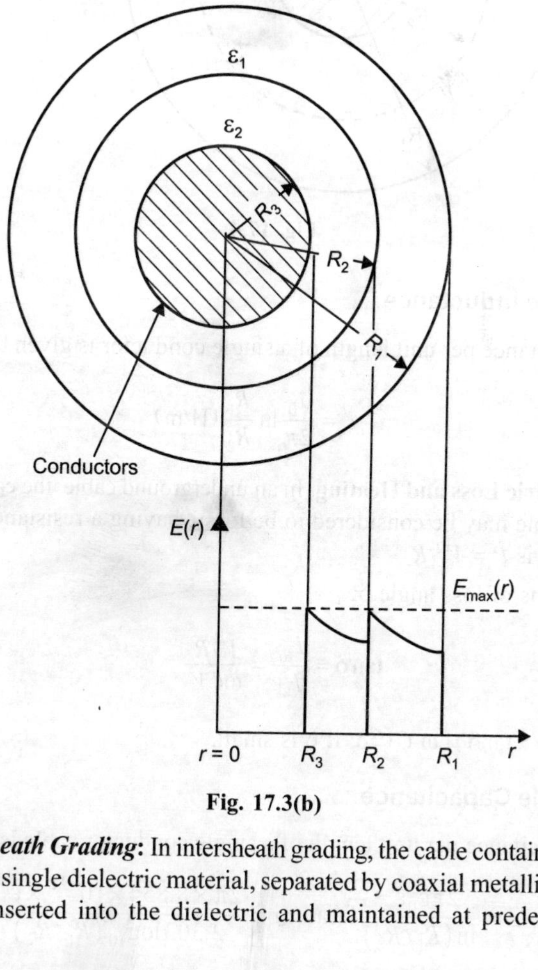

Fig. 17.3(b)

Intersheath Grading: In intersheath grading, the cable contains several layers of a single dielectric material, separated by coaxial metallic sheaths that are inserted into the dielectric and maintained at predetermined voltages. A cable with one such intersheath is shown in Fig. 17.4.

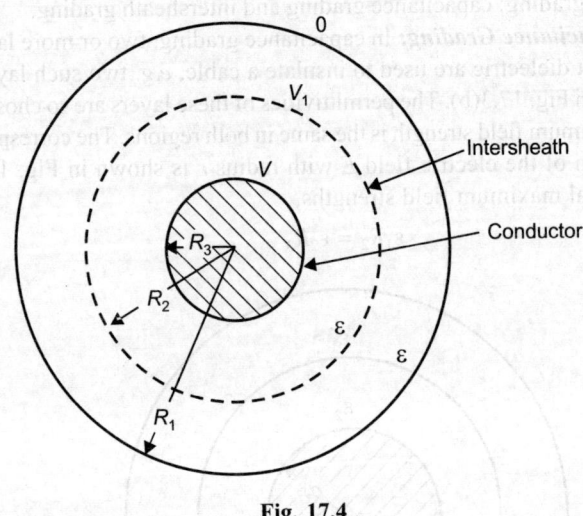

Fig. 17.4

20. Cable Inductance

The inductance per unit length of a single conductor is given by

$$L = \frac{\mu_0}{2\pi} \ln \frac{R_1}{R_2} (\text{H/m})$$

Dielectric Loss and Heating. In an underground cable, the capacitance of the cable may be considered to be lossy, having a resistance R_i, The loss in R_i is $P = V^2/R_i$

In terms of loss angle δ;

$$\tan \delta = \frac{I_{Ri}}{I_C} = \frac{V/R}{\omega CV}$$

$P = \omega C \, V^2 \tan\delta \cong \omega \, CV^2\delta$; if δ is small.

21. Cable Capacitance

The capacitance per unit length of a single conductor cable is given by

$$C = \frac{2\pi\varepsilon_0\varepsilon_r}{\ln(R_1/R_2)} \text{ farad/m} = \frac{2\pi \times 8.854 \times 10^{-12} \times \varepsilon_r}{2.303 \log_{10}(R_1/R_2)} \text{ F/m}$$

$$= \frac{\varepsilon_r \times 10^{-9}}{41.4 \log_{10}(R_1/R_2)} \text{ F/m}$$

In a three core cable, the empirical formula for capacitance of each core to neutral C_n is given by

$$\frac{0.03\varepsilon_r}{\log_e \left[1 + (T + t/d)(3.84 - 1.7)(t/T) + 0.52\left(t^2/T^2\right) \right]} \mu F/km.$$

where d = diameter of conductor, t = thickness of belt insulation and T = thickness of conductor insulation.

22. Important Corona Terms

Critical Disruptive Voltage: It is the minimum phase to neutral voltage at which corona occurs. Consider two conductors of radii r cm and spaced d cm apart. If V is the phase to neutral potential, then potential gradient at the conductor surface is given by

$$g_{max} = \frac{V}{r \log_e (d/r)} \text{ volt/cm}$$

In order that corona is formed, the value of g_{max} must be made equal to the breakdown strength of air. The breakdown strength of air at 76 cm pressure and temperature of 25 °C is 30 kV/cm (*peak*) or 21.2 kV/cm (rms) and is denoted by g_0. If V_C is the phase-neutral potential required under these conditions, then

$$g_0 = \frac{V_c}{r \log_e d/r}$$

where, g_0 = breakdown strength of air at 76 cm of mercury and 25 °C
= 30 kV/cm (peak) or 21.2 kV/cm (RMS)

$\therefore$ Critical disruptive voltage $V_C = g_0\, r \log_e \dfrac{d}{r}$

The above expression for disruptive voltage is under standard conditions, *i.e.* at 76 cm of Hg and 25 °C. However, if these conditions vary, the air density also changes, thus altering the value of g_0. The value of g_0 is directly proportional to air density. Thus, the breakdown strength of air at a barometric pressure of b cm of mercury and temperature of t °C becomes δg_0, where

$$\delta = \text{air density factor} = \frac{3.92b}{273 + t}$$

Under standard conditions, the value of $\delta = 1$

$\therefore$ Critical disruptive voltage, $V_C = g_0 \delta r \log_e \dfrac{d}{r}$

Correction must also made for the surface condition of conductor. This is accounted for by multiplying the above expression by irregularity factor m_0.

$\therefore$ Critical disruptive voltage, $V_c = m_0 g_0 \delta r \log_e \dfrac{d}{r}$ kV/phase

where, m_0 = 1 for polished conductors

          = 0.98 to 0.92 for dirty conductors

          = 0.87 to 0.8 for stranded conductors.

23. Visual Critical Voltage

It is the minimum phase-neutral voltage at which corona glow appears all along the line conductors.

It has been seen that in case of parallel conductors, the corona glow does not begin at the disruptive voltage V_C but at a higher voltage V_V, called *visual critical voltage*. The phase-neutral effective value of visual critical voltage is given by the following empirical formula:

$$V_V = m_v g_0 \delta r \left(1 + \frac{0.3}{\sqrt{\delta r}}\right) \log_e \frac{d}{r} \text{ kV/phase}$$

where m_v is another irregularity factor having a value of 1.0 for polished conductors and 0.72 to 0.82 for rough conductors.

24. Power Loss Due to Corona

Formation of corona is always accompanied by energy loss which is dissipated in the form of light, heat, sound and chemical action. When disruptive voltage is exceeded, the power loss due to corona is given by:

$$P = 244 \left(\frac{f + 25}{\delta}\right) \sqrt{\frac{r}{d}} \left(V - V_c\right)^2 \times 10^{-5} \text{ kW/km/phase}$$

25. Caclulation of Sag

Consider a conductor suspended between two equilevel supports A and B in still air as shown in Fig. 17.5, where 0 is the lowest point on the conductor which is the middle of span.

Let $L =$ Length of span (m), w = weight/m of conductor (kg)

$T =$ tension in conductor (kg). Assume the sag is very small so curved length of conductor A, B is equal to length of span. Consider a point P on conductor at a distance x metre from the mid point O. The following two external forces are acting on the portion OP of the conductor.

(i) the weight of conductor of x metres, *i.e.* w.x. kg is acting at $x/2$ metres from point P giving anticlockwise moment.

(ii) the tension T acting at point O at y metres from point P giving clockwise moment.

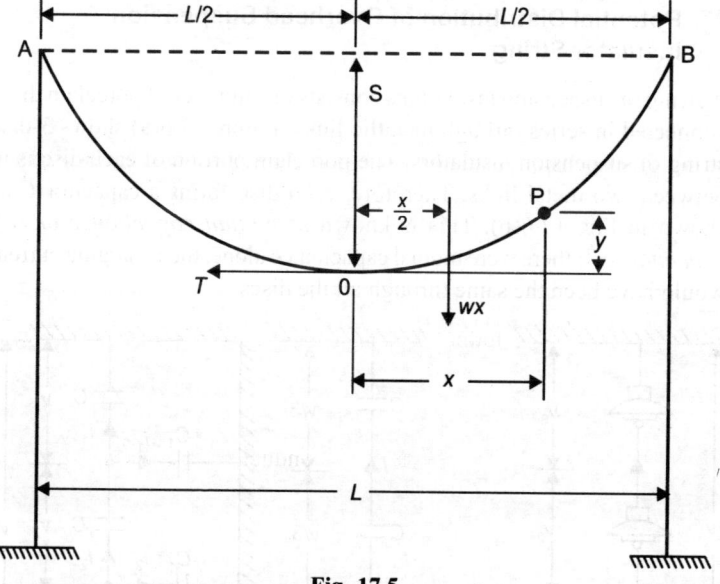

Fig. 17.5

26. Effect of Wind and Ice Loading

In still air, sag develops due to the weight of conductor only. In actual practice, a conductor may have ice coating and simultaneously subjects to wind pressure. The ice load (w_i) acts vertically downward, *i.e.* in the same direction as the weight of conductor, whereas, the wind load (w_w) acts horizontally, *i.e.* perpendicular to the weight of conductor.

Resultant weight of conductor per unit length,

$$w_r = \sqrt{(w + w_i) + (w_w)^2}$$

where w = weight of conductor per unit length,

w_i = weight of ice per unit length = density of ice × volume of ice per unit length.

$$= \text{density of ice} \times \frac{\pi}{4}(D^2 - d^2) \times 1$$

$$= \text{density of ice} \times [(d + 2t)^2 - d^2] \times 1$$

w_w = wind load per unit length

Sag under this condition, $S = \dfrac{w_r L^2}{8T}$

27. Potential Distribution of Overhead Suspension Insulator String

A string of suspension insulations consists of a number of porcelain discs connected in series through metallic links. Figure 17.6(a) shows 3-disc string of suspension insulators. The porcelain portion of each disc is in between two metal links. Therefore, each disc forms a capacitor C as shown in Fig. 17.6(b). This is known as *mutual capacitance or self capacitance.* If there were mutual capacitance alone, then charging current would have been the same through all the discs.

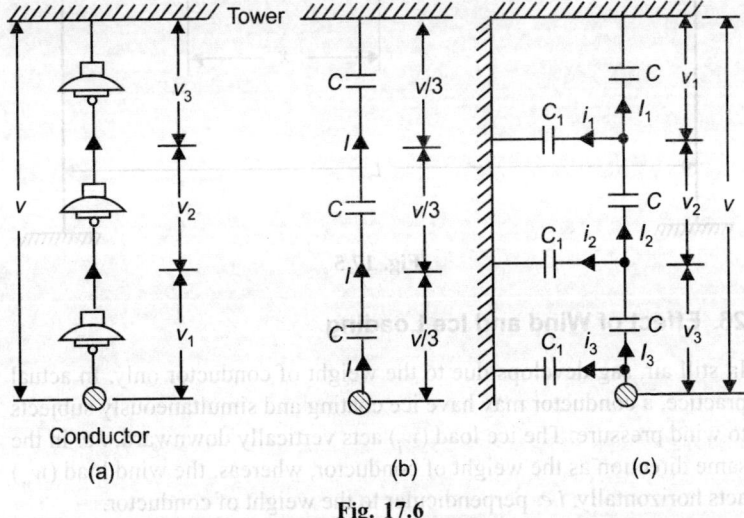

Fig. 17.6

Figure 17.6(c) shows the equivalent circuit for a 3-disc string. Let that self capacitance of each disc is C and shunt capacitance $C_1 = KC$.

The voltage across each unit is V_1, V_2 and V_3 are given by

$$\therefore \qquad V_3 = V_1[1 + 3K + K^2]$$

Voltage between conductor and earth (*i.e.* tower)

$$V = V_1 + V_2 + V_3 = V_1 + V_1(1 + K) + V_1(1 + 3K + K^2)$$
$$= V_1(3 + 4K + K^2) = V_1(1 + K)(3 + K)$$

$$\therefore \qquad \text{Voltage across top unit, } V_1 = \frac{V}{(1+K)(3+K)}$$

28. String Efficiency

The ratio of voltage across the whole string to the product of number of discs and the voltage across the disc nearest to the conductor is known as *string efficiency*, *i.e.*

$$\text{String efficiency} = \frac{\text{Voltage across the string}}{n \times \text{voltage across disc nearest to the conductor}}$$

where n = number of discs in the string.

Chapter 18

Protection

1. Terms Related to Fault Calculations

(i) Percentage Reactance: It is defined as follows:

$$\% \text{ reactance} = \frac{\text{Reactance} \times (\text{kVA})}{10(\text{kV})^2}$$

(ii) Percentage reactance at base kVA. Since different plants in any system have different kVA ratings, it is necessary to convert all reactances to a common kVA rating base.

$$\% \text{ Reactance at base kVA} = \frac{\text{Base kVA}}{\text{Plant kVA}} \times \text{reactance at plant kVA}$$

2. Per Unit

Per unit impedance of electrical equipment

$$\varepsilon_z = (\text{I}.Z)/V \, p.u$$

Per unit impedance is given by:

$$\text{Base } Z = \frac{[\text{Base kV}]^2}{[\text{Base kVA}]} \times 1000$$

$$p.u. \, Z = \frac{\text{Ohmic } Z}{\text{Base } Z} = \frac{Z \cdot [\text{Base kVA}]}{[\text{Base kV}]^2 \times 1000}$$

$$\%Z = Z \cdot \frac{[\text{Base kVA}]}{10[\text{Base kVA}]^2}$$

Change of Base *p.u.* $Z_{\text{old}} = p.u. Z_{\text{old}} \times \left[\dfrac{\text{kVA Base (New)}}{\text{kVA Base (Old)}} \right] \times \left[\dfrac{\text{kV Base (Old)}}{\text{kV Base (New)}} \right]^2$

Percentage Impedance = 100 p.u. impedance

When all the reactances have been referred to the same base kVA, then total impedance or reactance upto the fault point is calculated. This may involve solution of series parallel circuits or star/delta conversion or *vice versa*.

$$\text{Short circuit kVA} = \left[\frac{100}{\%Z} \times \text{kVA} \right]$$

$$\text{Short circuit current (rms)} = \frac{\text{Short circuit kVA}}{\sqrt{3} \times \text{kV (line)}} \text{ amp}$$

3. Unsymmetrical Currents

The unbalanced phase currents in a three-phase system in terms of symmetrical components are:

$$I_R = I_{R1} + I_{R2} + I_{RO}; \ I_Y = I_{Y1} + I_{Y2} + I_{Y0}; \ I_B = I_{B1} + I_{B2} + I_{B0}$$

(i) Zero Sequence Current

$$\therefore \qquad I_0 = \frac{1}{3}(I_R + I_Y + I_B)$$

(ii) Positive Sequence Current

$$\therefore \qquad I_{R1} = \frac{1}{3}\left(I_R + aI_Y + a^2 I_B\right)$$

Omitting the subscript *R*, we have $I_1 = \frac{1}{3}\left(I_R + aI_Y + a^2 I_B\right)$

(iii) Negative Sequence Current

$$I_{R2} = \frac{1}{3}\left(I_R + a^2 I_Y + aI_B\right)$$

4. Restriking Voltage

$$v(t) = \sqrt{2}\, V\left[1 - \cos(\omega_0 t)\right] = V_m\left[1 - \cos\left(\frac{1}{LC}\right)t\right]$$

The restriking voltage (*v*) can thus rise to a maximum value of $2V_m$, where V_m is the peak value of the system recovery voltage. If a resistance

R_s be connected across the contacts of the circuit breaker, the surge will be critically damped when $R_s = \dfrac{1}{2}\sqrt{L/C}$ and this offers an important method of reducing the severity of transient.

5. Current Chopping

Current chopping arises with air blast circuit breakers which operate on the same air pressure and velocity for all values of interrupted current. On low current interruption, the breaker tends to open the circuit before the current natural zero and the electromagnetic energy present is rapidly converted to electrostatic energy.

$$\frac{1}{2}Li_0^2 = \frac{1}{2}Cv^2 \text{ joules} \qquad \therefore \quad v = i_0\sqrt{L/C}$$

If resistance and time is also included, this equation becomes

$$v = i_0\sqrt{L/C}\, e^{-\alpha t}\sin\omega_0 t, \qquad \omega_0 = 1/\sqrt{LC}$$

High transient voltages may be set-up on opening a highly inductive current such as transformer on no load.

6. Resistance Switching

A deliberate connection of resistance in parallel with contact space is called *resistance switching*. (Fig. 18.1). Resistance switching is resorted to circuit breakers having high post zero resistance of contact space.

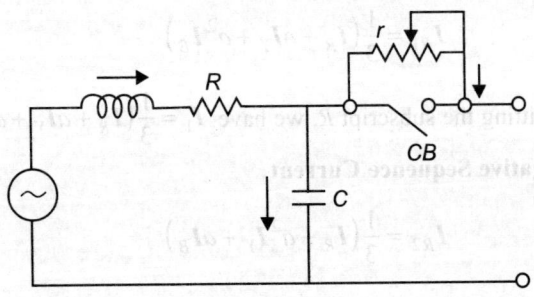

Fig. 18.1

The magnitude of resistance for resistance switching is given by $r = \dfrac{1}{2}\sqrt{\dfrac{L}{C}}$.

7. Active Recovery Voltage

The maximum value of the system phase voltage $= \sqrt{2}V_{ph} = V_m$

Active recovery voltage, $V_{AR} = K_1 K_2 K_3 V_m \sin \phi$

where K_1 is called the *demagnetizing factor* due to which the recovery voltage will be less than the system voltage

or $K_1 = \dfrac{\text{Recovery voltage}}{\text{System voltage}}$

K_2 is a condition factor, *i.e.* it depends on the condition whether the symmetrical fault is grounded or not, *i.e.* its value is either 1 or 1.5.

and K_3 is a factor equal to 1 if the active recovery voltage between phase and neutral is to be obtained and its value is $\sqrt{3}$ if the active recovery voltage between the two lines is required.

8. Circuit Breaker Ratings

  (i) Breaking capacity
 (ii) Making capacity
(iii) Short time capacity

The *breaking* (*rupturing*) *capacity* of circuit breakers is expressed in kVA or MVA, being obtained from the product of short circuit current and recovery voltage.

The *symmetrical breaking current* of a circuit breaker is the current which it will interrupt at a power factor of 0.15 for ratings upto 500 MVA and at a power factor of 0.3 for ratings of 750 MVA or upwards with a recovery voltage of 95 per cent normal voltage.

The *asymmetrical breaking current* of a circuit breaker is the current which it will interrupt when there is asymmetry in one of the phases.

Asymmetrical breaking current,

$$I = \sqrt{I_{DC}^2 + I_{AC}^2} = \sqrt{(0.5I_m)^2 + \left(\frac{I_m}{\sqrt{2}}\right)^2} = 0.866\, I_m \cong 1.25\, I_{AC}$$

which means that under these conditions, rated asymmetrical breaking current will be equal to 1.25 times the rated symmetrical breaking current for a circuit breaker.

The (*peak*) *making current* of a circuit breaker is the peak value of the maximum current wave (including the DC component) in the first cycle of the current after the circuit is closed by the circuit breaker.

Rated (peak) making capacity $\cong \sqrt{2} \times 1.8$, i.e. $2.55 \times$ rated symmetrical breaking capacity.

The *short-time rated current* is the current that can safely be applied, with the circuit breaker in normal condition for 3 seconds if the ratio of symmetrical breaking current to normal current is less than 40.

Short-time opening duty: Any circuit breaker must be capable of the following duty:

$$B - 3' - MB - 3' - MB$$

B denotes breaking operation, MB denotes a making operation followed by a breaking operation without intentional time delay while 3 denotes a 3 minute time interval.

Rating of the motor = RMS hp

$$= \sqrt{\frac{1}{\text{Time for 1 cycle}} \int (hp)^2 \, dt} = \sqrt{\frac{\sum (hp)^2 \times \text{time}}{\text{Time for 1 cycle}}}$$

9. Inertia Constant and Swing Equation

The *per-unit constant H* is defined as the kinetic energy stored in the rotating parts of the machine at *synchronous speed* per unit mega volt ampere (MVA) rating of the machine.

$$H = \frac{\text{Stored energy in mega joules}}{\text{Rating in MVA}}$$

Thus, if G is the *MVA* rating of the machine, then $GH = \frac{1}{2} J \omega_s^2$ where J is the polar moment of inertia of all rotating parts in kilogram $\cdot$ m^2 and ω_s is the angular synchronous velocity in electrical radians per second. If M is the corresponding angular momentum, then

$$M = J \, \omega_s$$

Since $\omega_s = 360°$ electrical degrees per second,

$$M = \frac{GH}{180f} \text{ MJ. sec/electrical degree}$$

$$= \frac{GH}{\pi f} \text{ megajoules.sec/radian}$$

where f is the frequency of rotation.

Swing Equation: If P_a is the accelerating or decelerating power, then

$$P_a = P_i - P_e$$

$$= M \frac{d^2\theta}{dt^2}$$

$$\frac{H}{180f} \cdot \frac{d^2\delta}{dt^2} = P_i - P_e$$

$$= P_a \text{ per unit}$$

10. Constant on a Common MVA Base

An inertia constant H_{mach} based on a machine's own MVA rating may be converted to a value H_{syst} relative to the system base S_{syst} with the formula

$$H_{syst} = H_{mach} = \frac{S_{mach}}{S_{syst}}$$

A convenient system base value is 100 MVA.

11. Equal Area Criterion

For stability, the area under the graph of accelerating power P_a vs δ shown in Fig. 18.2 must be zero for some value of δ; that is, the positive area A_1 under the graph must be equal to the negative area A_2. This criterion is therefore known as the *equal area criterion* for stability.

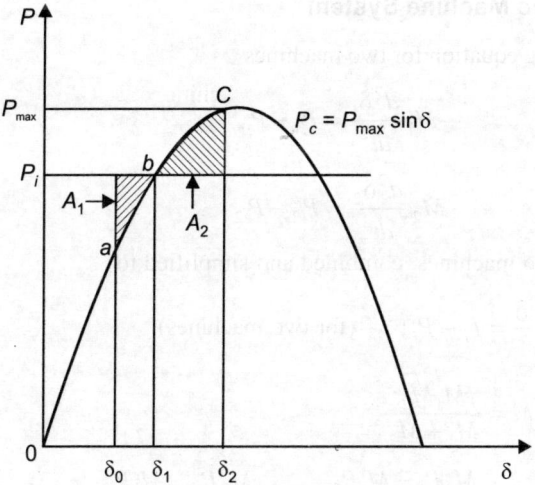

Fig. 18.2

12. Critical Clearing Angle

Figure 18.3 shows a power angle curve A before a fault, curve B during the fault, and curve C after the fault such that $A = P_{max} \sin\delta$, before the fault, $B = r_1 P_{max} \sin\delta$ during the fault and $C = r_2 P_{max} \sin\delta$ after the fault with $r_1 < r_2$ for stability.

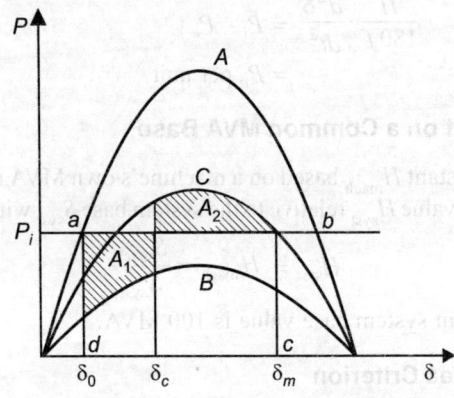

Fig. 18.3

$$\cos\delta_c = \frac{1}{r_2 - r_1}\left[(\delta_m - \delta_0)\sin\delta_0 - r_1\cos\delta_0 + r_2\cos\delta_m\right]$$

13. A Two Machine System

The Swing equation for two machines

$$M_1\frac{d^2\delta_1}{dt^2} = P_{i1} - P_{e1}$$

$$M_2\frac{d^2\delta_2}{dt^2} = P_{i2} - P_{e2}$$

For two machines, combined and simplified to

$$M\frac{d^2\delta}{dt^2} = P_i - P_e ; \quad \text{(for two machines)}$$

where $\quad M = \dfrac{M_1 M_2}{M_1 + M_2}$,

$$P_i = \frac{M_2 P_{i1} - M_1 P_{i2}}{M_1 + M_2}, \quad P_e = \frac{M_2 P_{e1} - M_1 P_{e2}}{M_1 + M_2}$$

Chapter 19

Induction and Dielectric Heating

1. Classification of Heating Methods

Heating Methods

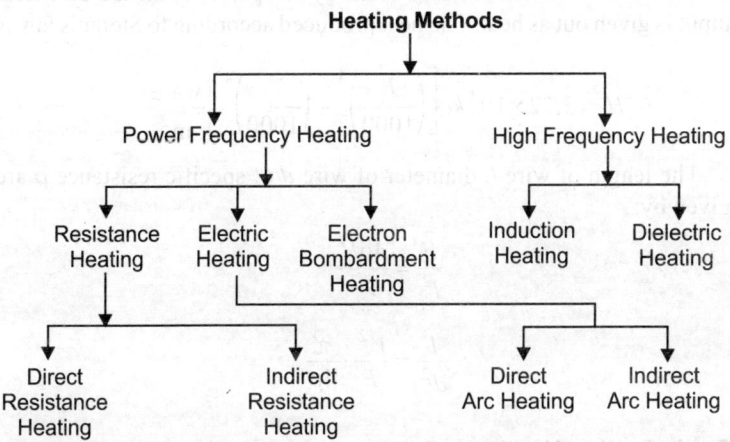

2. Resistance Heating

When electric current passes through a resistance, power loss takes place therein, which appears in the form of heat.

$$\text{Power loss} = I^2R = IV = \frac{V^2}{R} \text{ watts}$$

where R is the effective resistance of the element.

173

3.   Different Types of Heating Materials

Types of Alloy	Composition	Max Temp. of Operation	Sp. Resistance at Room Temp.	Sp. Gravity
Nickel–Copper	45% Ni 55% Cu	400 °C	49 μohm/cm^3	8.88
Nickel–Chromium Iron	60% Ni, 16% Cr, 24% Fe	950 °C	110 μohm/cm^3	8.28
Nickel–Chromium	80%Ni, 20% Cr	1150 °C	109 μohm/cm^3	8.36
Iron-Chromium Aluminium	65–75% Fe, 20–30% Cr, 5% Al	1150– 1350 °C	140 μohm/cm^3	7.2

4.   Design of Resistance Heating Elements

When the element has reached a steady temperature, all the electrical input is given out as heat. The heat produced according to Stefan's law is

$$H = 5.72 \times 10^4 \, ke \left[\left(\frac{T_1}{1000} \right)^4 - \left(\frac{T_2}{1000} \right)^4 \right] \frac{\text{watts}}{\text{m}^2}$$

The length of wire *l*, diameter of wire *d* of specific resistance ρ are given by:

$$\frac{d}{l^2} = \frac{4\rho H}{V^2}$$

$$\frac{l}{d^2} = \frac{V^2}{P} \cdot \frac{\pi}{4\rho}$$

5.   Induction Heating

Depth of penetration

$$\delta = \frac{1}{2\pi \sqrt{10^{-9} \, \mu_r \sigma f}} \, \text{cm}$$

where μ_r = relative magnetic permeability of the material

σ = conductivity, mhos/cm^3 and

f = frequency of the input current (Hz)

For induction heating of cylindrical bars, the surface power density is given by

$$P = \frac{4H_0}{\sigma\delta} \text{ watt/cm}^2$$

The desirable frequency for a given work piece should be more than the critical frequency given by

$$f_c \cong \frac{2}{u_r \sigma a^2}$$

where H_0 is magnetic field intensity at the surface of piece of metal, σ the conductivity, μ_r the relative permeability of the material and 'a' is the cross-sectional area.

6. Dielectric Heating

Consider an infinite small cube of dielectric material placed in an AC electrostatic field as shown in Fig. 19.1, the potential difference between two faces of the cube is V. Let the side of the cube be 'a'. The capacitance of the cube is given by

$$C = \varepsilon_0 \varepsilon_r \frac{\text{Area of the face}}{\text{Thickness}} = \varepsilon_0 \varepsilon_r \frac{a^2}{a} = \varepsilon_0 \varepsilon_r a \text{ farad}$$

where ε_r is the relative permittivity of the material. The resulting current makes an angle θ with the voltage V across the capacitor. θ is slightly less than 90°. The small in-phase component I is given by

$$I \cos\theta = \frac{|V|}{Z_C} \cos\theta$$

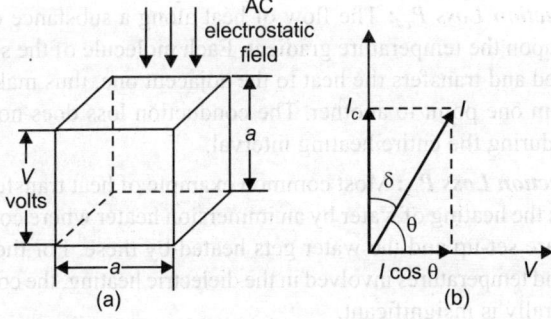

Fig. 19.1

where Z_C is the impedance of elementary capacitor and nearly equals to $1/2\pi fC$.

The power dissipated in the cube $= 2\pi f \cdot \varepsilon_0 \varepsilon_r a \cdot V^2$. (pf)

Power per unit volume $P_v = 2\pi f \cdot \varepsilon_0 \varepsilon_r (pf) \, v^2$ watt/m^3

where v is the applied voltage per unit length. The product of dielectric constant ε_r and power factor (pf), *i.e.* $(\varepsilon_r \cdot pf)$ is called the *loss factor*.

$$f_c = \frac{14.31}{a\sqrt{\varepsilon_r}} \text{ MHz, where } a \text{ is the radius of material.}$$

7. Thermal Losses in Dielectric Heating

The total power P_i delivered to a dielectric piece is given by

$$P_i = P_s + P_{cd} + P_{cv} + P_{cp} + P_{cr} \text{ watts}$$

where P_s is the specific heat power in watts or thermal power

P_{cd} is the conduction loss in watts

P_{cv} is the convection loss in watts

P_{cp} is the power in watts required to change the physical state of material if needed

P_{cr} is the radiation loss in watts

Thermal Power P_s: This is the power required to raise the temperature of a given dielectric material to the desired final temperature within a specified heating time.

Specific heat power: $P_s = 17.6 \ mc \ \Delta T$;

where m is the mass, c is the specific heat and ΔT is the temperature rise.

Conduction Loss P_{cd}: The flow of heat along a substance or object depends upon the temperature gradient. Each molecule of the substance gets heated and transfers the heat to the adjacent one, thus making heat travel from one point to another. The conduction loss does not remain constant during the entire heating interval.

Convection Loss P_{cv}: Most common example of heat transfer by this method is the heating of water by an immersion heater where convection currents are set-up and the water gets heated by these. For the heating periods and temperatures involved in the dielectric heating, the convection loss generally is insignificant.

Radiation Loss P_{cr}: Heat reaches the object from the source without heating the medium in between, (Stefan's law of heat radiation).

$$\text{Heat dissipated} = 5.72 \times 10^4 \, ke\left[\left(\frac{T_1}{1000}\right)^4 - \left(\frac{T_2}{1000}\right)^4\right] \text{ watts/m}^2$$

where k is a constant called *radiating efficiency* whose value is 1 for single element and 0.5 to 0.8 for several elements placed side by side.

ε is emissivity equal to 1 for black body and 0.9 for resistance heating elements.

T_1 is the temperature of the source in °K observed.

T_2 is the temperature of object absorbing the heat in °K.

Chapter 20

Power Electronics

1. Single Phase Separately Excited DC Motor Drive

The basic circuit for single phase separately excited DC motor drive is shown in Fig. 20.1. The armature voltage is controlled by a semiconverter or full converter and the field circuit is fed from the AC supply through a diode bridge. The motor current cannot reverse due to thyristors in the converters. If semiconverters are used, the average DC output voltage E_a is always positive. Thus, power flow $E_a I_a$ is always positive, *i.e.* from AC supply to DC load. In semiconverters, freewheeling action takes place when thyristor blocks.

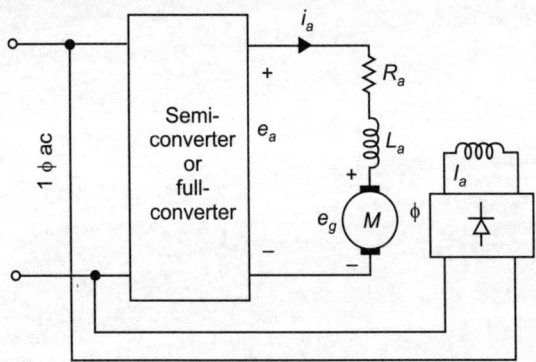

Fig. 20.1

2. Basic Equations

The armature circuit of the DC motor is represented by its back voltage e_g, armature resistance R_a, armature inductance L_a as shown in Fig. 20.1.

Back voltage	$e_g = K_a \Phi N$
Average back voltage	$E_g = K_a \Phi N$
Developed torque	$t = K_a \Phi i_a$
Average developed torque	$T_{av} = K_a \Phi I_a$
The armature circuit voltage equation	$e_a = R_a i_a + L_a \dfrac{di_a}{dt} + e_g$
In terms of average values	$E_a = R_a I_a + E_g$
The average speed	$N = \dfrac{E_a - I_a R_a}{K_a \Phi}$

3. Continuous Armature Current

For continuous armature current condition, circuits are shown in Figs 20.2 and 20.3 for semiconverter and full converter systems respectively.

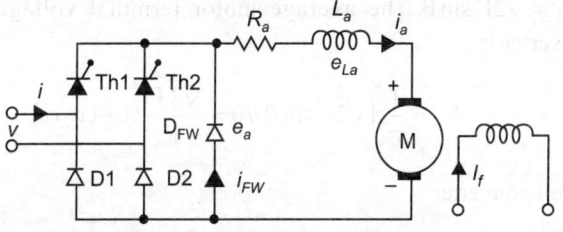

Fig. 20.2

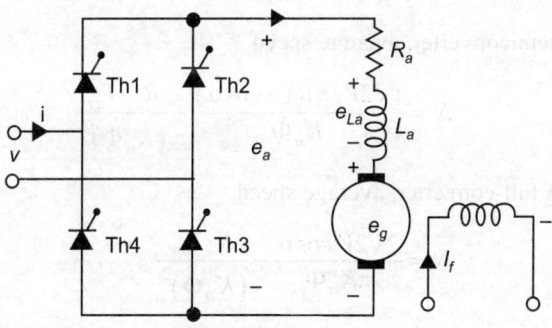

Fig. 20.3

In the semiconverter system, shown in Fig. 20.2, thyristor Th1 triggered at an angle α and Th2 at an angle $\alpha + \pi$ with respect to supply voltage v.

In the full converter systems shown in Fig. 20.3, thyristors Th1 and Th3 are simultaneously triggered at α and thyristors Th2 and Th4 are triggered at $\pi + \alpha$.

4. Torque Speed Characteristics

For a semiconverter with freewheeling action, the circuit equations are:

$$e_a = v = R_a i_a + L_a \frac{di_a}{dt} + e_g \qquad \alpha < \omega t < \pi$$

$$e_a = 0 = R_a i_a + L_a \frac{di_a}{dt} + e_g \qquad \pi < \omega t < \pi + \alpha$$

The armature circuit equation for a full converter is

$$e_a = v = R_a i_a + L_a \frac{di_a}{dt} + e_g \qquad \alpha < \omega t < \pi + \alpha$$

Let $v = \sqrt{2}V \sin\theta$. The average motor terminal voltage with a semiconverter is

$$E_a = \frac{1}{\pi} \int_{\alpha}^{\pi} \sqrt{2}V \sin\theta\, d\theta = \frac{\sqrt{2}V}{\pi}(1 + \cos\alpha)$$

With a full converter

$$E_a = \frac{1}{\pi} \int_{\alpha}^{\pi+\alpha} \sqrt{2}V \sin\theta\, d\theta = \frac{2\sqrt{2}V}{\pi}\cos\alpha$$

With semiconverter, average speed

$$N = \frac{(\sqrt{2}V/\pi)(1 + \cos\alpha)}{K_a\Phi} - \frac{R_a \cdot T_{av}}{(K_a\Phi)^2}$$

With a full-converter, average speed

$$N = \frac{2\sqrt{2}V\cos\alpha}{\pi K_a\Phi} - \frac{R \cdot T_{av}}{(K_a\Phi)^2}$$

The first term in the above equation represents theoretical no load speed. The second term represents speed drop produced by armature current I_a and hence torque T_{av}. The theoretical no load speed is varied by the firing angle α.

5. Single Phase DC Series Motor Drives

Basic Equations: The armature circuit resistance R_a and inductance L_a include the resistance and inductance of the series field winding. Back emf is

$$E_g = K_a \, \Phi n.$$

The flux Φ has two components. One component, say Φ_a, is produced by the armature current flowing through the series field winding. The other component, say Φ_{res}, is due to the residual magnetism. The latter is small and can be assumed constant.

$$\Phi = \Phi_a + \Phi_{res}$$

If magnetic linearity is assumed, $\Phi_a = K_f i_a$

$$e_g = K_a \, (K_f i_a + \Phi_{res}) \, n = K_a K_f i_a n + K_a \Phi_{res} n$$

$$= K_{af} i_a n + K_{res} \, n$$

The back voltage due to residual magnetism is very small and is proportional to speed. The back voltage due to the flux produced by the armature current is the major voltage and it is present when both i_a and n are present. Average back voltage is

$$E_g = K_{af} N I_a + K_{res} N$$

Developed torque $\qquad t = K_a \, \Phi i_a$

If the flux Φ_{res} is neglected, then $t \approx K_{af} \cdot i_a^2$

Therefore, for either direction of current, torque is developed in the same direction. Hence in the series motor, speed reversal can be achieved by reversing either the field winding or the armature terminals but not both.

Average torque = $T_{av} = K_{af} i_a^2 \, |_{\text{average}} = K_{af} I_{av}^2$

The armature circuit voltage equation is given by

$$e_a = R_a i_a + L_a \frac{di_a}{dt} + e_g$$

In terms of average quantities, $E_a = R_a \, I_a + E_g$

The circuit diagrams for semiconverter and full converter series motor drives are shown in Figs 20.4 and 20.5.

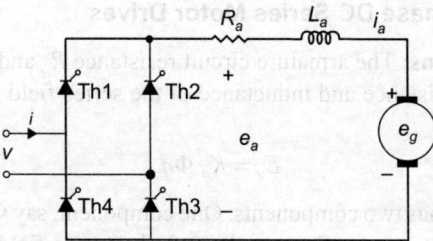

Fig. 20.4

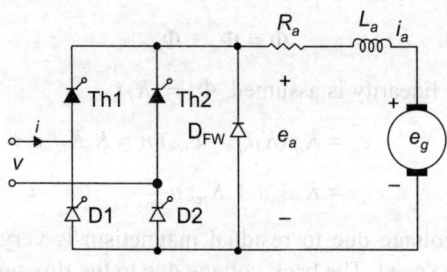

Fig. 20.5

6. Continuous Motor Current

In phase controlled series motor drives, the motor current is mostly continuous. In terms of average voltages, the voltage equations are as follows:

For semiconverter systems, $E_a = \dfrac{\sqrt{2}V}{\pi}(1+\cos\alpha)$

$$= R_a I_a + K_{af}I_a N + K_{res} N$$

For full converter systems, $E_a = \dfrac{2\sqrt{2}V}{\pi}\cos\alpha$

$$= R_a I_a + K_{af}I_a N + K_{res} N$$

For semiconverter system, the average speed is

$$N = \frac{\left(\sqrt{2}V/\pi\right)(1+\cos\alpha)-R_a I_a}{K_{af}I_a + K_{res}}$$

For full converter system $N = \dfrac{\left(2\sqrt{2}V/\pi\right)\cos\alpha - R_a I_a}{K_{af}I_a + K_{res}}$

If the ripple in the motor current is neglected, then $I_a \sim I_{av}$

Average torque $\quad T_{av} = K_{af} I_{av}^2 \sim K_{af} I_a^2$

For semiconverter system,

$$T_{av} = K_{af} \left[\frac{(\sqrt{2}V/\pi)(1 + \cos\alpha) - K_{res}N}{R_a + K_{ef}N} \right]^2$$

For full converter system,

$$T_{av} = K_{af} \left[\frac{(2\sqrt{2}V/\pi)(\cos\alpha) - K_{res}N}{R_a + K_{ef}N} \right]^2$$

7. Three Phase Drives

Figures 20.6 and 20.7 show a three phase semiconverter and full converter drive circuits

Basic equations for semiconverters

$$E_a(\alpha) = I_a R_a + E_g = I_a R_a + K_a \Phi N$$

$$N = \frac{E_a(\alpha) - R_a I_a}{K_a \Phi}.$$

Let supply phase voltage be $v_A = \sqrt{2}V \sin\omega t$

$$v_B = \sqrt{2}V \sin\left(\omega t - \frac{2\pi}{3}\right)$$

$$v_0 = \sqrt{2}V \sin\left(\omega t + \frac{2\pi}{3}\right)$$

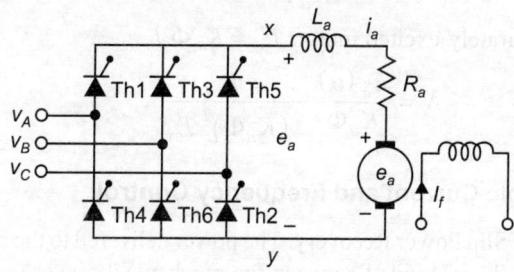

Fig. 20.6

If the motor currents is continuous,

$$E_a(\alpha) = \frac{3}{2\pi} \int_{\pi/6+\alpha}^{\pi/6+a+2\pi/3} (v_a - v_c) \, d(\omega t)$$

$$= \frac{3\sqrt{6}V}{2\pi} (1 + \cos\alpha)$$

The displacement angle ϕ_1 increases as the firing angle increases. The input power factor will thus decrease as firing angle increases.

8. Basic Equations for Full Converter

$$E_a(\alpha) = \frac{3}{\pi} \int_{\pi/6+\alpha}^{\pi/6+\alpha+\pi/3} (v_A - v_B) \, d(\omega t)$$

$$= \frac{3\sqrt{6}V}{2\pi} (1 + \cos\alpha)$$

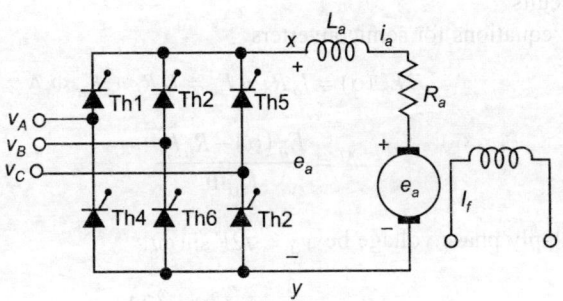

Fig. 20.7

The average speed $= \dfrac{E_a(\alpha) - R_a I_a}{K_a \Phi}$

In a separately excited motor, $T_{av} = K_a \, \Phi \, I_a$.

So, $N = \dfrac{E_a(\alpha)}{K_a \Phi} - \dfrac{R_a}{(K_a \Phi)^2 \, T_{av}}$

9. Variable Current and Frequency Control

Principle of Slip Power Recovery: The power delivered to the rotor across the air gap P_{ag} is divided between the mechanical power output P_{mech}, and the rotor copper loss, P_{Cu} given by $s \, P_{ag}$.

and $$P_{mech} = (1 - s) P_{ag}$$

Also $$P_{ag} = T\omega_s$$

where T is the electromagnetic torque developed by the motor, and ω_s is the synchronous angular velocity.

10. Induction Motor Control by Choppers

The speed of wound rotor induction motor can be varied by varying the rotor resistance. The rotor resistance can be varied steplessly by using a chopper circuit shown in Fig. 20.8.

The time T_{ON} during which thyristor is conducting, R_1 is the effective rotor resistance and the time T_{OFF} during which thyristor is blocked and an effective resistance $R_1 + R_2$ is in the rotor circuit. Assuming sufficiently high value of chopper frequency. Thus, for the total time $T = T_{ON} + T_{OFF}$, the effective value of resistance is given by

$$R = \frac{R_1 T_{ON} + (R_1 + R_2) T_{OFF}}{T_{ON} + T_{OFF}}$$

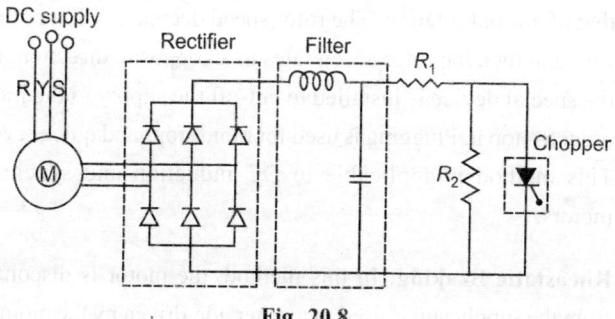

Fig. 20.8

Hence, the rotor resistance can be varied by varying time $T_{ON} + T_{OFF}$.

Chapter 21

Braking

1. Electric Braking

(i) **Plugging:** It requires the reversal of connections to the armature of the motor so that it develops a torque in opposite direction to that of normal rotation. The rotor speed decreases till it reduces to zero and then the rotor accelerates in an opposite direction, unless the special device is installed to cut-off the supply to the motor in order to stop it. Plugging is used for rapid stops and quick reversals. This method is applicable to DC induction and synchronous motors.

(ii) **Rheostatic Braking:** In this method, the motor is disconnected from the supply and is used as a generator, driven by the momentum of the equipment to be braked, the electric energy so generated is dissipated as heat the external resistors. This method can be used for DC, induction and *synchronous motors.*

(iii) **Regenerative Braking:** With regenerative braking, the motor continues to be connected to the supply but it made to act as a generator. Kinetic energy of the drive is converted into electrical energy part of which is fed back to the supply system and part of it is dissipated as heat in the winding of the machine.

2. Fundamental Formulae in Dynamics

Linear Motion	Angular Motion
Mass $= \dfrac{W}{g}$ kg	Moment of inertia J in kg $\cdot$ m^2
Linear acceleration $a = f_t \times \dfrac{g}{W}$ m/sec	Angular acceleration $\alpha = \dfrac{d\omega}{dt}$ radians/sec.
Linear velocity, $v = \dfrac{ds}{dt}$ m/sec	Angular velocity, $\omega = 2\pi\dfrac{dn}{dt}$ radians/sec. or $n = 1/2\pi \int \omega.dt$
Linear distance, $s = \int v.dt$ m.	Angular motion, $n = \int \dfrac{W}{2\pi}d\theta$
Tractive force $f_t = \dfrac{W}{g}a$ kg	Torque, $T_m = \dfrac{J}{g}\alpha$ kg $\cdot$ m

3. Time Calculations

Angular retardation

$$\alpha_r = \frac{T_B g}{J}$$

$$t = \frac{J}{Kg}\log_e \frac{K\omega_i + T_M}{T_M}$$

Speed:
$$n = \frac{1}{2\pi K}\left[\frac{J}{gK}(K\omega_i + T_M)(1 - e^{gK/jt}) - T_M \times t \right]$$

4. Standard Ratings for Motors

The ratings of a motor may either be continuous, continuous maximum or short time. ISI specification has defined these ratings as given below:

(i) *Continuous rating*: This is a rating or the output of a motor which is capable of delivering continuously without exceeding the maximum permissible temperature rise. This term also indicates that the motor is capable of delivering 25% overload for 2 hours.

(ii) *Continuous maximum rating*: This term is similar to the continuous rating with the only difference is that the motor is not capable of running on overload. The term is employed for motors having capacity more than 2.5 hp per rpm.

(iii) *Short time rating*: This term denotes the output of a motor for a short and specified period but without exceeding the specified temperature rise.

Choice of Motor

Assuming heating to be proportional to load.

Rating of the motor = (Rms) HP

$$= \sqrt{\frac{1}{\text{time for 1 cycle}} \int (HP)^2 \, dt}$$

$$= \sqrt{\frac{\sum (HP)^2 \times \text{time}}{\text{time for 1 cycle}}}$$

Chapter 22

Electronics

1. Electron Emission

Richardson Dushman Equation: The emission current density:

$$J_s = AT^2 \, e^{-b/T} \text{ amp/m}^2$$

where,

J_s = emission current density, T = absolute temperature of emitter in °K

A = constant depending upon the type of the emitter, amp/m^2/°K

b = a constant for the emitter, e = natural logarithmic base

The value of b is constant for a metal and is given by $\phi \, e/k$

where ϕ is work function of emitter

e = electron charge = 1.602×10^{-19} coulomb

k = Boltzmann's constant = 1.38×10^{-23} J/°K

$$\therefore \quad b = \frac{\phi + 1.602 \times 10^{-19}}{1.38 \times 10^{-23}} = 11600 \, \phi \, °K$$

or

$$J_s = AT^2 e^{-\frac{11600}{T}\phi}$$

Commonly used Thermionic Emitters

S.No.	Emitter	Work Function	Operating Temperature	Emission Efficiency
1.	Tungsten	4.52 eV	2327°C	4 mA/W
2.	Thoriated tungsten	2.63 eV	1700°C	60 mA/W
3.	Oxide coated	1.1 eV	750°C	200 mA/W

2. Child's Law

In the space charge limited region, the plate current $I_b = K E_b^{3/2}$, where K is a constant whose value depends on the shape of the electrode and geometry of the tube and E_b is the plate voltage.

Relationship between u, r_p and g_m

Amplification factor = Plate resistance × mutual conductance

or $\qquad\qquad\qquad u = r_p \times g_m$

3. Comparison of Valve Constants

S. No.	Particular	Triode	Tetrode	Pentode
1.	Amplification factor (u)	10 to 100	Range around 500	1000 to 5000
2.	Plate resistance (r_p)	300 to 1000 kohm	70 to 1000 kohm	0.5 to 20 μ-ohm
3.	Transconductance (g_m)	About 2500 μ-mho	About 1000 μ-ohm	1000 to 9000 μ-ohm

4. Semiconductors

Current in Semiconductor

In a semiconductor, both holes and electrons contribute to electrical conduction. With an applied electric field E, the expression for current density is

$$J = (n\mu_n + p\mu_p)\, eE = \sigma E$$

where n and p are the concentrations of electrons and holes (number/m^3) and μ_n and μ_p are the corresponding mobilities, e is the electronic charge. Conductivity depends on the number of charge carriers and their mobility; for a semiconductor, conductivity is

$$\sigma = (n\mu_n + p\mu_p)\, e$$

In a pure semiconductor, the number of holes is just equal to the number of conduction electrons or $n = p = n_i$,
where n_i is the intrinsic concentration.

In a doped semiconductor,

$$np = n_i^2$$

In other words, the product of electron and hole concentrations is a constant; if one is increased (by doping), the other must decrease. If the doping concentration is nonuniform, and it is possible to have charge motion by the mechanism called *diffusion*. The *diffusion current* is proportional to the concentration gradient dn/dx. the diffusion current density due to electrons is given by

$$J_n = eD_n \frac{dn}{dx}$$

where D_n is the diffusion constant for electrons (m^2/s).

The diffusion current density due to nonuniform concentrations of randomly moving electrons and holes is

$$J = J_n + J_p = eD_n \frac{dn}{dx} - eD_p \frac{dp}{dx}$$

5. Semiconductor Diodes

A semiconductor diode conducts forward current with a small forward voltage drop across the device, simulating a closed switch. The relationship between the forward current and forward voltage is a good approximation given by the Shockely diode equation,

$$i = I_s[e^x - 1],$$

where $$x = \frac{eV}{kT}$$

and I_s is the leakage current through the diode, e is electronic charge, k is Boltzman's constant, T is the temperature of diode, and V is the voltage across the diode.

6. Transistors

Transistor Connections: (a) Common base configuration (b) Common emitter configuration (c) Common collector configuration.

Common base configuration (CB)

Current amplification factor (α): It is the ratio of change in collector current to the change in emitter current at a constant collector base voltage V_{cn}, i.e.

$$\alpha = \frac{\Delta t_c}{\Delta I_E} \text{ at constant } V_{CB}$$

Collector current: $I_C = \alpha I_E + I_{CBO}$, where I_{CBO} is the leakage collector current.

Input resistance: It is the ratio of change in emitter base voltage (ΔV_{EB} to the resulting change in emitter current (ΔI_E) at constant collector base voltage (V_{CB}), *i.e.* input resistance $r_i = \dfrac{v_{EB}}{I_F}$ at constant. (V_{CB}).

Output resistance: It is the ratio of change in collector base voltage (ΔV_{CB}) to the change in collector current (ΔI_c) at constant emitter current, *i.e.* output resistance, or, $r_o = \Delta V_{CB}/\Delta I_C$ at constant I_E.

Common Emitter Configuration (CE)

Base Current Amplification Factor (β): It is the ratio of change in collector current (ΔI_E) to the change in base current (ΔI_B)

i.e. $$\beta = \frac{\Delta I_C}{\Delta I_B}$$

Relation between β and α: $\beta = \dfrac{\alpha}{1-\alpha}$

Collector current $I_C = \beta I_B + I_{CEO}$

where I_{CEO} is the collector emitter current with base open.

Emitter current $I_E = (\beta + I) I_B + I_{CEO}$

Input resistance: It is the ratio of change in base emitter voltage (ΔV_{BE}) to the change in base current (ΔI_B) at constant V_{CE}.

Input resistance $r_i = \Delta V_{BE}/\Delta I_B$ at constant V_{CE}.

Common Collector Configuration (CC)

Current Amplification Factor (γ): It is the ratio of change in emitter current (ΔI_E) to the change in base current (ΔI_B), i.e. $\gamma = \Delta I_E/\Delta I_B$

Relation between γ and α: $\gamma = \dfrac{1}{1-\alpha}$

Collector current $\qquad I_C = \alpha I_E + I_{CBO}$

Emitter current $\qquad I_E = (\beta + 1) I_\beta + (\beta + 1) I_{CBO}$

$\qquad\qquad\qquad\qquad = I_B/(1 - \alpha) + I_{CBO}/(1 - \alpha)$

7. Comparison of Transistor Configurations

S. No.	Emitter	Work Function	Operating Temperature	Emission Efficiency
1.	Input resistance	Low (about 100 ohm)	Low (about 750 ohm)	Very high (about 750 kohm)
2.	Output resistance	Very high (about 450 kohm)	High (about 45 kohm)	Low (about 50 ohm)
3.	Voltage gain	about 150	about 500	less than 1
4.	For frequency applications	For high frequency applications	For audio frequency applications	For impedance matching

8. Field Effect Transistors

In a *junction field effect transistor* (JFET), the width of depletion layers controls conductance. For the JFET, the drain current in the constant-current–region is

$$i_{DS} = I_{DSS}(1 - V_{GS}/V_p)^2,$$

where i_{DS} is the drain current in the constant–current region, I_{DSS} is the value of i_{DS} with gate shorted to source, and V_p is the pinch-off voltage.

For an enhancement MOSFET, the transfer characteristic is

$$i_{DS} = K(V_{GS} - V_T)^2$$

where K is a device parameter and V_T is the turn-on or threshold voltage.

9. Bipolar Junction Transistor (BJT)

A bipolar junction transistor consists of two *pn* junctions in close proximity normally, the emitter junction is forward biased, the collector is reverse biased. In common-base operation, the collector current

$$i_C = -\alpha i_E + I_{CBO}, \quad \text{where } \alpha \cong 1$$

where I_{CBO} is the collector cut-off current and α is the forward current–transfer ratio. In common emitter operation, a small base current controls the relatively larger collector current to achieve current amplification.

$$i_C = \beta i_B + I_{CEO}, \quad \text{where } \beta = \frac{\alpha}{1-\alpha}$$

where i_B is the base current and I_{CEO} is the collector cut-off current in the common emitter configuration.

10. Ebers–Moll Equations

$$i_E = I_{ES}(e^q V_{EB/KT} - 1) - \alpha_R I_{CS}(e^q V_{CB/KT} - 1)$$

$$i_C = \alpha I_{ES}(e^q V_{EB/KT} - 1) - I_{CS}(e^q {}_{CB/KT} - 1)$$

Hybrid Model

The pair of equations for the two-port network is shown in Fig. 22.1.

$$v_1 = h_{11}i_1 + h_{12}v_2$$

$$i_1 = h_{21}i_1 + h_{22}v_2$$

For a transistor connected in Common–Emitter (CE) configuration, base is the input terminal and collector is the output terminal with respect to the terminal.

Therefore,

$$v_{be} = h_{11}i_b + h_{12}v_{ce}$$

$$i_c = h_{21}i_b + h_{22}v_{ce}$$

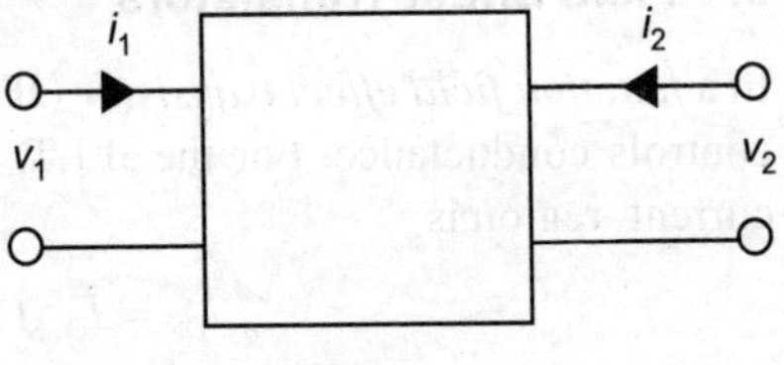

Fig. 22.1

From these equations, the h-parameters can be defined as follows:

$$h_{11} = h_{ie} = \frac{v_{be}}{i_b}\Big|_{v_{ce}=0} = \text{Short-circuit input impedance}$$

$$h_{12} = h_{re} = \frac{v_{be}}{v_{ce}}\Big|_{i_b=0} = \text{Open-circuit reverse voltage gain}$$

$$h_{21} = h_{fe} = \frac{i_c}{i_b}\Big|_{v_{ce}=0} = \text{Short-circuit forward current gain}$$

$$h_{22} = h_{oe} = \frac{i_c}{i_{ce}}\Big|_{i_b=0} = \text{Open-circuit output admittance}$$

Thus, the equations become

$$v_{be} = h_{ie}\,i_b + h_{re}\,v_{ce}$$

$$i_c = h_{fe}i_b + h_{oe}v_{ce}$$

11. Performance of Linear Circuit in *h*-Parameters and for Common Emitter Configuration

(i) **Input impedance,** $Z_m = h_{11} - h_{12}\, h_{21} \Big/ \Big(h_{22} + \dfrac{1}{r_2} \Big)$

$Z_i \approx h_{11}$ (if h_{12} or r_L is very small)

(ii) **Current gain, A** $\quad = \dfrac{h_{21}}{1 + h_{22} + r_L} \approx h_{21}\,(h_{22} \cdot r_L \ll 1)$

$$A_i = \dfrac{h_{fe}}{1 + h_{oe} r_L}$$

(iii) **Voltage gain,** $\quad A_v = \dfrac{-h_{21}}{Z_i \Big(h_{22} + \dfrac{1}{r_L} \Big)}$

or $\qquad\qquad A_v = \dfrac{-h_{fe}}{Z_i \big(h_{poe} + 1/r_L \big)}$

12. Integrating Circuit

$$v_0 = \frac{1}{RC} \int_0^t v_i\, dt$$

The output voltage is proportional to the integral of the input voltage under the assumption that R_{Ci} is very large, the time as compared to period of input wave [Fig. 22.2].

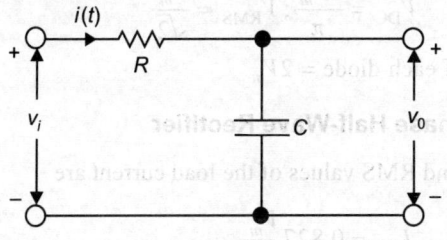

Fig. 22.2

13. Differentiator

By making the RC, time constant small with respect to the interval of any applied voltage, the output voltage

$$v_0 = iR \cong RC \frac{dv_i}{dt}$$

Thus, with small value of time constant RC, the output voltage will be proportional to the derivative of the input voltage and the circuit is known as **differentiating circuit** (Fig. 22.3).

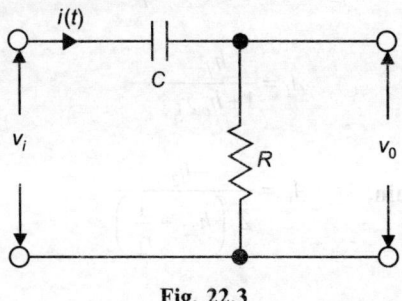

Fig. 22.3

14. Half-Wave Rectifier

Average and RMS values of load voltage for input $V_i = V_m \sin \omega t$ are:

$$V_{DC} = \frac{V_m}{\pi}; \ V_{RMS} = \frac{V_m}{2}$$

Peak inverse voltage (PIV) of the diode $= V_m$

15. Centre-tap Type Full Wave Rectifier

The average and RMS values of the load voltage are given by

$$V_{DC} = \frac{2V_m}{\pi}; \ V_{RMS} = \frac{V_m}{\sqrt{2}}$$

The PIV of each diode $= 2V_m$

16. Three Phase Half-Wave Rectifier

The average and RMS values of the load current are

$$I_{DC} = 0.827 \frac{V_m}{R_L}$$

$$I_{\text{RMS}} = \sqrt{\frac{3}{2\pi} \int_{-\pi/3}^{+\pi/3} \frac{V_m^2 \cos^2 \omega t}{R_L^2} d\omega t}$$

$$= \frac{0.838}{R_L} V_m$$

17. The m-Phase Rectifier

The relations of an *m* phase rectifier:

$$I_{\text{DC}} = \frac{m}{2\pi} \int_{-\pi/m}^{+\pi/m} \frac{V_m \cos \omega t}{R_L} d\omega t = \frac{V_m}{R_L} \cdot \frac{m}{\pi} \sin \frac{\pi}{m}$$

$$I_{\text{RMS}} = \left[\frac{\frac{1}{2\pi} \int_{-\pi/m}^{\pi n/m} V_m^2 \cos^2 \omega t \, d\, \omega t}{R_L^2} \right]$$

$$= \frac{V_m}{R_L} \frac{1}{2\pi} \left(\frac{\pi}{m} + \sin \frac{\pi}{m} \cos \frac{\pi}{m} \right)$$

$$\text{Ripple factor, } \gamma = \frac{I_{\text{RMS}}}{I_{\text{DC}}}$$

$$= \frac{1}{3\sqrt{2}} \cdot \frac{R_L}{\omega L}$$

$$= 0.236 \frac{R_L}{\omega_L}$$

18. Comparison of Rectifiers

S. No.	Particular	Half-Wave	Centre-Top	Bridge Type
1.	No. of diodes	1	2	4
2.	Transformer required	No	Yes	No
3.	Maximum efficiency	40.6%	81.2%	81.2%
4.	Ripple factor	1.21	0.48	0.48
5.	Output frequency	f	$2f$	$2f$
6.	Peak inverse voltage (PIV)	V_m	$2V_m$	V_m

19. SCR With Resistive Load

The DC current through load *d* is given by

$$I_{DC} = \frac{1}{2\pi}\int_{\phi}^{\pi}\frac{V_m \sin \omega t}{R_L}\,d\omega t$$

$$= \frac{V_m}{2\pi R_L}(1+\cos\phi)$$

20. Speed Control of DC Motors

The voltage equation is

$$V_{DC} - I_a R_a - E_b = 0$$

and
$$E_b = K_1\,\Phi_f N$$

Therefore
$$N = \frac{V_{DC} - I_a R_a}{K_1 \Phi_F}$$

Chapter 23

Digital Logic

1. Digital Logic: Postulates

a. An operator '+' is defined such that if $c = a + b$ then $c \in X$, for every pair of elements $a, b \in X$. This operation replaces the 'union' (U) operation of set theory, and in switching theory, it is known as OR operation.

b. There exists as element '0' in X such that $a + 0 = a$ for every elements $a \in X$.

c. It holds commutative law true:

(i) $a + b = b + a$

(ii) $a \cdot b = b \cdot a$

d. It holds the distributive law true:

$$a \cdot (b + c) = a \cdot b + a \cdot c$$

e. The complementary set $\bar{a}$, for every element $a \in X$, is defined as

(i) $a + \bar{a} = 1$

(ii) $a \cdot \bar{a} = 0$

f. There are at least two elements $a, b \in X$ such that $a \neq b$.

Based on these postulates, one can define now AND, OR and NOT operations. For example, the absorption law stated earlier, now can be written as

$$A \cdot (A + B) = A$$

$$A + A \cdot B = A.$$

DeMorgan's Theorem (Inverse of Boolean Function)

It states that, "to obtain the inverse of any Boolean function invert all variables and replace all ORs by ANDs and all ANDs by ORs". The first De Morgan theorem says that a NOR gate $\overline{(A + B)}$ is equivalent to an AND gate with NOT circuits in the inputs $(\overline{A} \cdot \overline{B})$. The second says that a NAND gate $(\overline{A} \cdot \overline{B})$ is equivalent to an OR gate with NOT circuits in the input $\overline{(AB)}$.

2. AND Gate

A *logic gate* a device that controls the flow of information, usually in the form of pulses. The symbol for an AND gate is shown in Fig. 23.1. A . B is read "A AND B." As indicated in the truth table, an output appears only when there are inputs at A AND B.

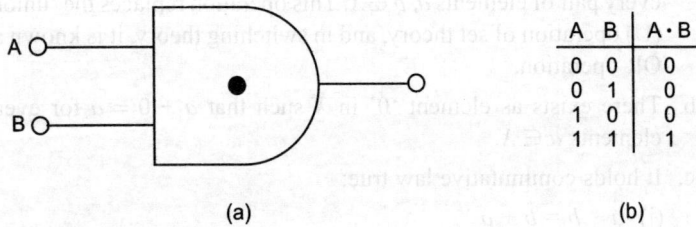

A	B	A·B
0	0	0
0	1	0
1	0	0
1	1	1

(a)   (b)

Fig. 23.1 (a) Symbol and (b) truth table for the AND gate

3. OR Gate

The symbol for an OR gate is shown in Fig. 23.2, where A + B is read "A OR B." as indicated in the truth table, the output is 1 in input A OR input B is 1. For no input, the output is zero (0).

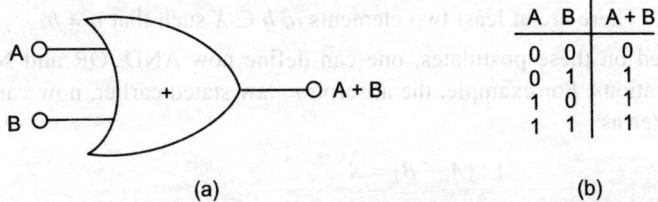

A	B	A + B
0	0	0
0	1	1
1	0	1
1	1	1

(a)   (b)

Fig. 23.2 A two-input OR gate (a) Symbol and (b) truth table

4. NOT Gate

The logic NOT is represented by the symbol in Fig. 23.3, where # is read "NOT A." As indicated in the truth table the NOT element is an *inverter*;

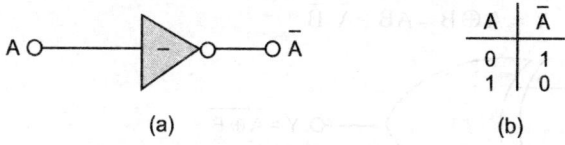

A	$\bar{A}$
0	1
1	0

(a) (b)

Fig. 23.3 A NOT gate (a) symbol and (b) truth table

the output is the *complement* of the single input.

5. NAND Gate

The NAND gate is defined by the truth table of Fig. 23.4. The circle on the NAND element symbol and the bar on the $\overline{A \cdot B}$ output indicate the inversion process.

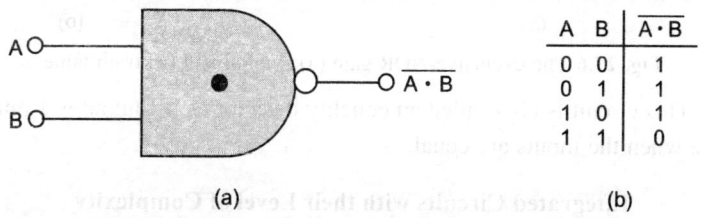

A	B	$\overline{A \cdot B}$
0	0	1
0	1	1
1	0	1
1	1	0

(a) (b)

Fig. 23.4 The NAND gate (a) symbol and (b) truth table

6. Exclusive-OR Gate

The Exclusive-OR operation is $(A + B)\overline{AB}$ shown in Fig. 23.5. The Exclusive-OR gate is used so frequently that it is represented by the special symbol $\oplus$ defined by $A \oplus B$.

$$A \times OR \quad B = A \oplus B = (A + B)\overline{AB}$$

A	B	$A \oplus B$
0	0	0
0	1	1
1	0	1
1	1	0

(a) (b)

Fig. 23.5 The Exclusive-OR gate (a) symbol and (b) truth table

7. Exclusive NOR Gate

The exclusive NOR gate, abbreviated as EX-NOR, operates exactly opposite to EX-OR gate.

$$Y = A \oplus B = AB + \overline{A} \cdot \overline{B}$$

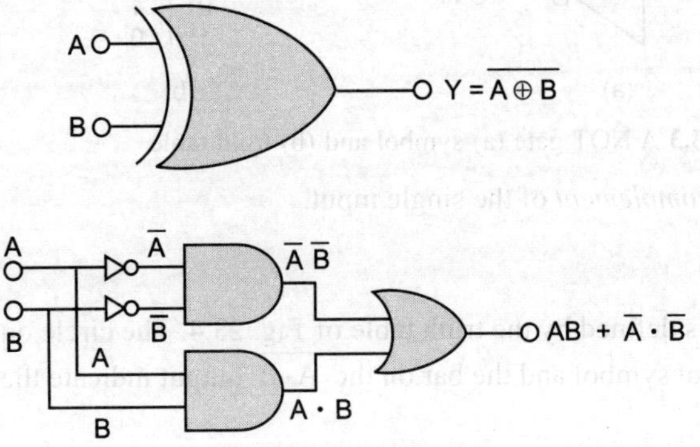

A	B	A⊕B
0	0	1
0	1	0
1	0	0
1	1	1

(a)   (b)

Fig. 23.6 The exclusive-NOR gate (a) symbol and (b) truth table

This circuit is also called an equality detector as its output is 1 only the when the inputs are equal.

Integrated Circuits with their Level of Complexity

Complexity	Approximate no. of gates per chip	Typical products
Small scale integration (SSI)	Less than 12	Logic gates, flip flops
Medium scale	12 to 99	Counters, multiplexers, adders
Large scale integration (MSI)	100 to 9999	8 bit microprocessors ROM, RAM
Very large scale integration (VLSI)	10,000 to 99,999	16 bit and 32 bit microprocessors, sophisticated computer peripherals
Ultra large scale integration	100,000 or more	64 bit microprocessors real-time image processing

Chapter 24

Communication Systems

1. Half-Power Bandwidth

The constancy of the magnitude $|H(j\omega)|$ of a system is specified by a parameter called *bandwidth B*, and is defined as the interval of positive frequencies over which $|H(\omega)|$ remains within 3 dB (with $1/\sqrt{2}$ in voltage or $\dfrac{1}{2}$ in power).

Amplitude Modulation

The equation of a general sinusoidal (carrier) signal can be written as

$$y(t) = a(t) \cos[\omega_c t + \phi(t)]$$

where $a(t)$ and $\phi(t)$ vary slowly compared to $\omega_c t$. The team $a(t)$ is called the envelope of the signal, ω_c is the carrier frequency, and $\phi(t)$ is the phase modulation of $y(t)$.

The modulating signal $x(t)$ provides an amplitude modulated carrier signal as

$$x(t) = kx(t) \cos \omega_c t$$

where $x(t) = 0$ and k is a constant.

Phase and Frequency Modulation

The modulating signal $x(t)$ can be used to modulate the frequency or phase of the carrier signal as

$$y(t) = A \cos \theta(T)$$

where A is a constant.

The relation between instantaneous angular rate $\omega(t)$ and $\phi(t)$ is

$$\theta(t) = \int_0^t \omega(T)\,dT + \theta_0$$

or

$$\omega(t) = \frac{d\delta}{dt}$$

Phase modulation is obtained when

$$\theta(t) = \omega_c\, t + k_1\, x(t) + \theta_0$$

and $x(t)$ is the modulating signal.

2. Measure of Information

The measure of information with an event A occuring probability P_A is

$$I_A = \log 1/P_A \text{ with } \log_2 \text{ (base 2)}$$

Entropy

The average information, called *entropy H* of a message is

$$H = \sum_{i=1}^n P_i \log_2 \frac{1}{P_i}$$

The Sampling Theory

A real-valued band limited signal having no spectral components above a frequency $B\,(H_z)$ is determined uniquely by its values at uniform intervals spaced on greater than $1/2B$ seconds apart.

For signal $x(t)$ with a Fourier transform $X(f)$, where $X(f)$ is assumed zero for $f \geq B$, the signal is recoverable from a sampling frequency f_s:

$$f_s \geq 2B$$

Only time functions which are continuous, single-valued and band limited are considered.

Theorem 1: If a time function $f(t)$ is continuous, no frequency components higher than W cps then the time function can be completely determined by specifying its ordinates at a series of points spaced every 1/2 W sec or less.

Theorem 2: If a time function $f(t)$ is composed of a band of frequencies displaced from zero with a bandwidth of W cycles, with highest frequency of f_2 then the minimum sampling rate is given by $2f_2/m$, where m is the largest integer not to exceed f_2/W. All higher sampling rates are not necessarily usable.

Theorem 3: If $F(t)$ represents the frequency spectrum of a time function, $f(t)$ which is zero everywhere except in the range $T_1 \le t \le T_2$ then $F(f)$ can be uniquely determined by specifying its values at a series of points spaced every $1/(T_2 - T_1)$ cycles.

Theorem 4: If a signal whose highest frequency is W cycles has been sampled at a rate of $2\,W$ samples/sec, and the samples are in the form of impulses whose area is proportional to the magnitude of the sample at that instant, the sampled signal may be reconstructed by passing the impulse train through an ideal-pass filter whose cut-off frequency is W cycles.

Moments: Consider the discrete random variable X which assume the possible values $x_1, x_2, x_3, x_4, ..., x_m$. In a sequence of n experiments, let the event x_1 occur n_1 times x_2 occur n_2 times etc. The arithmetic average of mean of x_k is

$$X_{avg} = \frac{n_1 x_1 + n_2 x_2 + ... + n_m x_m}{n}$$

or $\qquad X_{avg} = \frac{1}{n} \sum_{k=1}^{m} x_k n_k$

Ergodic Processes: A random process is defined as ergodic, if all types of ensemble average are interchangeable with the corresponding time average. If the process is ergodic, then in general

$$\langle x^n(t) \rangle = \overline{X}^n, \text{ i.e. } \int_{-\infty}^{\infty} x^n p(x)\, dx = \underset{T \to \infty}{\text{Lt}} \frac{1}{2T} \int_{-T}^{T} x^n(t)\, dt$$

Autocorrelation Functions: The autocorrelation function of a random process is defined as

$$R(t_1,\, t_2) = E(X\,Y) \iint x\,y\,p(x, y)\,dx dy$$

where X and Y are random variables at times t_1 and t_2 respectively.

Spectral Density: Spectral density expresses the average power in a wave from (on a 1 ohm basis) as a function of frequency. Spectral density, $G(f)$ the average power or mean square voltage in a frequency interval Δf is simply $G(f)\,\Delta f$ and the mean square

$$\int_{-\infty}^{\infty} G(f)\, df$$

Discrete Channel: The ability of a discrete channel to transmit information may be measured in terms of a number of bits per unit time which is may transmit.

The capacity C of a discrete channel is given by

$$\lim_{T \to \alpha} \frac{\log N(T)}{T}$$

where $N(T)$ is the number of allowed signals in a duration T.

The channel capacity for this type of channel then follows as

$$C = \lim_{T \to \alpha} \frac{\log 2^{nT}}{T}$$

or simply $C = n$ bits/sec.

The Continuous System: A number of communication system utilize a continuous source and thus continuous use of channel amplitude, phase and frequency modulation are to mention just a few, where continuous use of channel is made. Measures of information for the discrete source having probabilities $P(1)$, $P(2)$,..., $P(i)$ is given by

$$H = -\sum_i P(i) \log P(i) \quad \text{bits/symbol}$$

Entropy of a sample point can be defined as

$$H = -\int_{-\infty}^{\infty} p(x) \log p(x) dx$$

where $p(x)$ is the probability of density function associated with the signal $x(t)$. Since the samples occur every $1/2$ W sec.

$$H'(x) = -2W \int_{-\infty}^{\infty} p(x) \log p(x) dx \quad \text{bits/sec}$$

The value of $H'(x)$ is a relative measure of information, subject to a coordinate system which may change.

Field Mapping

A field map is an electromagnetic road map, provides an overall physical picture and important insights about electric and magnetic field distributions.

Field maps can be produced by many different methods. The map of the microstrip transmission line of Fig. 24.1 was drawn graphically in 30 minutes. A graphical map is very useful as a check on solutions by other methods.

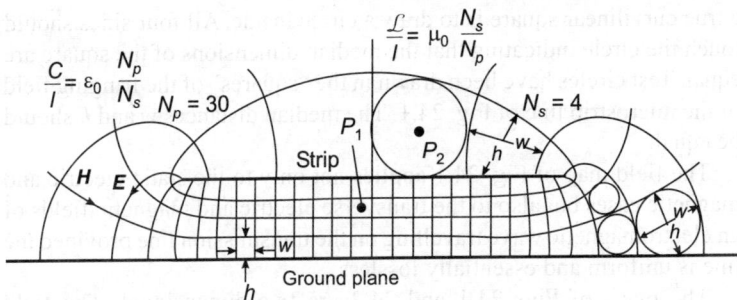

Fig. 24.1 Cross-section of microstrip transmission line mapped into field cells with test circles ($w = h$ for all cells, $\varepsilon_r = 1$ everywhere)

All solutions including graphical, are solutions of Laplace's equations. Such a solution is unique, it is the one and only solution.

Graphical Solution

Field maps become especially useful when divided into curvilinear squares. A *curvilinear square* is a four-sided area (with field and equipotentials intersecting at right angles) that tends to yield true squares as the area is subdivided into smaller area. Thus, the large area in Fig. 24.2 is a *curvilinear square* while the small one are true squares.

The microstrip transmission line of Fig. 24.2 is mapped into curvilinear squares or field cells. A simple way to determine whether a "square" is

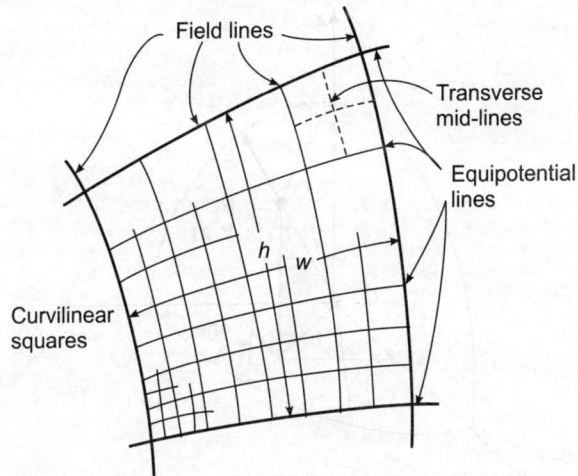

Fig. 24.2 Curvilinear square partially subdivided into smaller curvilinear squares

a true curvilinear square is to draw a circle inside. All four sides should touch the circle indicating that the median dimensions of the square are equal. Test circles have been drawn in the "squares" of the fringing field of the microstrip line of Fig. 24.1. The median distances w and h should be equal.

The field map of Fig. 24.1 applies not only to the static electric and magnetic cases but also to the transverse electric and magnetic fields of an electromagnetic wave travelling on the transmission line provided the line is uniform and essentially lossless.

The maps of Figs 24.1 and 24.2 are two-dimensional. The field configuration is the same for all planes perpendicular to a uniform transmission line. All such 2-dimensional geometries can be mapped into curvilinear squares.

Full Vector Notation

Referring to Fig. 24.3, the field E in rectangular coordinates at a point (x, y, z), and at a distance r from the origin is given by

$$E = \underbrace{\frac{Q_1}{4\pi\varepsilon_0} \frac{1}{\left(x^2 + y^2 + z^2\right)}}_{\text{magnitude}} \underbrace{\frac{x\,\hat{x} + y\,\hat{y} + z\,\hat{z}}{\left(x^2 + y^2 + z^2\right)^{1/2}}}_{\text{unit vector}}$$

$$= \frac{Q_1\left(x\,\hat{x} + y\,\hat{y} + z\,\hat{z}\right)}{4\pi\varepsilon_0\left(x^2 + y^2 + z^2\right)^{3/2}} \; \text{N} \cdot \text{C}^{-1} \qquad \ldots(24.1)$$

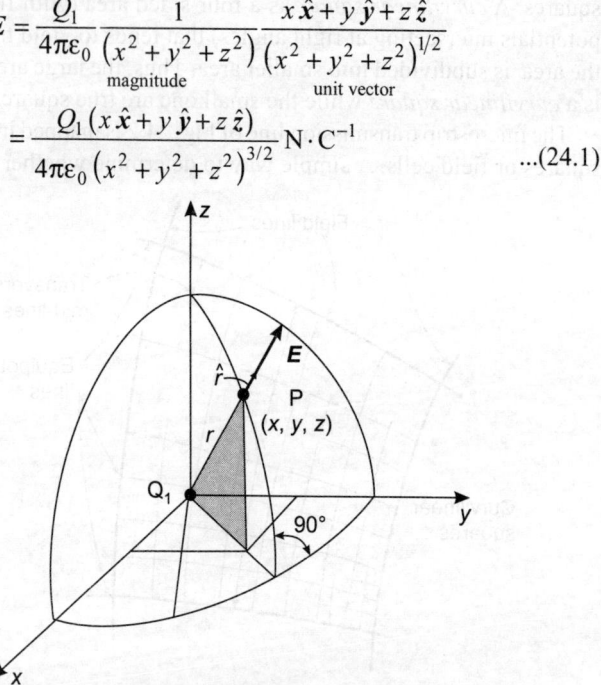

Fig. 24.3 Electric field at a point $P\,(x, y, z)$ from charge Q_1 at the origin

The field *E* at a point *P*(*x*, *y*, *z*) from a charge Q_1 at (x_1, y_1, z_1) (*not* at the origin) is given by

$$E = \frac{Q_1}{4\pi\varepsilon_0} \frac{(x - x_1)\hat{x} + (y - y_1)\hat{y} + (z - z_1)\hat{z}}{\left[(x - x_1)^2 + (y - y_1)^2 + (z - z_1)^2 \right]^{3/2}} \quad ...(24.2)$$

or, referring to Fig. 24.4, we have

$$E = \frac{Q_1}{4\pi\varepsilon_0 \left| R_p - R_1 \right|^2} \frac{R_p - R_1}{\left| R_p - R_1 \right|} = \frac{Q_1}{4\pi\varepsilon_0} \frac{\left(R_p - R_1 \right)}{\left| R_p - R_1 \right|^3} = \frac{Q_1}{4\pi\varepsilon_0} \frac{R_{1p}}{\left| R_{1p} \right|^3}$$

$$...(24.3)$$

If the charge is at the origin ($x_1 = y_1 = z_1 = 0$), Eq. (24.2) reduces to Eq. (24.1).

Example: *Electric Field.* Find the electric field at point (0, 2, 3) due to a charge $Q_1 = 10$ nC at (1, 0, 1). Dimensions in meter, $\varepsilon_r = 1$.

Solution. From Figs 24.4 and (24.3),

$$\mathbf{R}_p = 0\hat{x} + 2\hat{y} + 3\hat{z}$$

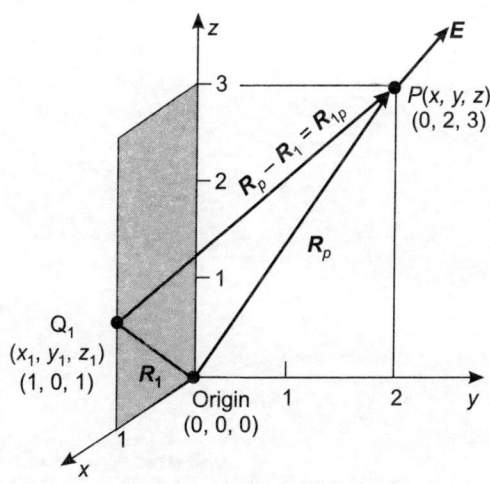

Fig. 24.4 Electric field at point *P* at (*x*, *y*, *z*) due to charge Q_1 at (x_1, y_1, z_1) (not at the origin) is given by Eqs (24.2) and (24.3).

APPENDIX A

Table A.1 Fundamental, mechanical, electrical, and magnetic units

Name of dimension or quantity	Symbol	Description	SI unit and abbreviation	Equivalent unit	Dimension
			Fundamental units		
Current	I	$\dfrac{\text{charge}}{\text{time}}$	ampere (A)	6.25×10^{18} electron charges per second $= \dfrac{C}{s}$	I
Length	L, l		metre (m)	$1,000\ \text{mm} = 100\ \text{cm}$	L
Mass	M, m		kilogram (kg)	$1,000\ \text{g}$	M
Time	T, t		second (s)	$\dfrac{1}{60}\ \text{min} = \dfrac{1}{3600}\ \text{h}$ $= \dfrac{1}{86400}\ \text{day}$	T

(Contd...)

Table A.1 Fundamental, mechanical, electrical, and magnetic units (*Contd.*)

Name of dimension or quantity	Symbol	Description	SI unit and abbreviation	Equivalent unit	Dimension
		Mechanical units			
Acceleration	a	$\dfrac{\text{velocity}}{\text{time}} = \dfrac{\text{length}}{\text{time}^2}$	$\dfrac{\text{metre}}{\text{second}^2}$ (m·s⁻²)		$\dfrac{L}{T^2}$
Area	A, a, s	length^2	metre^2 (m^2)		L^2
Energy or work	W	force × length = power × time	joule (J)	N·m = W·s = V·C $= 10^7$ ergs $= 10^8$ dynes·mm	$\dfrac{ML^2}{T^2}$
Energy density	w	$\dfrac{\text{energy}}{\text{volume}}$	$\dfrac{\text{joule}}{\text{meter}^3}$ (J·m⁻³)	$\dfrac{1}{100}\dfrac{\text{erg}}{\text{mm}^3}$	$\dfrac{M}{LT^2}$
Force	F	mass × acceleration	newton (N)	$\dfrac{\text{kg·m}}{\text{s}^2} = \dfrac{\text{J}}{\text{m}}$ $= 10^5$ dynes	$\dfrac{ML}{T^2}$

Quantity	Symbol	Definition	SI unit	Unit equivalence	Dimension
Frequency	f	$\dfrac{\text{cycles}}{\text{time}}$	hertz (Hz)	$\dfrac{\text{cycle}}{\text{s}}$	$\dfrac{1}{T}$
Impedance	Z	$\dfrac{\text{force}}{\text{mass} \times \text{velocity}}$	$\dfrac{\text{newton} \cdot \text{second}}{\text{kilogram} \cdot \text{metre}}$	$\dfrac{\text{N} \cdot \text{s}}{\text{kg} \cdot \text{m}}$	$\dfrac{1}{T}$
Length	L, l		metre (m)	1,000 mm = 100 cm	L
Mass	M, m		kilogram (kg)	1,000 g	M
Moment (torque)		force × length	newton · metre (N · m)	$\dfrac{\text{kg} \cdot \text{m}^2}{\text{s}^2} = \text{J}$	$\dfrac{ML^2}{T^2}$
Momentum	mv	mass × velocity = force × time = $\dfrac{\text{energy}}{\text{velocity}}$	newton · second (N · s)	$\dfrac{\text{kg} \cdot \text{m}}{\text{s}} = \dfrac{\text{J} \cdot \text{s}}{\text{m}}$	$\dfrac{ML}{T}$

(Contd...)

Table A.1 Fundamental, mechanical, electrical, and magnetic units (*Contd.*)

Name of dimension or quantity	Symbol	Description	SI unit and abbreviation	Equivalent unit	Dimension
Period	T	$\dfrac{1}{\text{frequency}}$	second (s)	T	T
Power	P	$\dfrac{\text{force} \times \text{length}}{\text{time}}$ $= \dfrac{\text{energy}}{\text{time}}$	watt (W)	$\dfrac{\text{J}}{\text{s}} = \dfrac{\text{N} \cdot \text{m}}{\text{s}}$ $= \dfrac{\text{kg} \cdot \text{m}^2}{\text{s}^3}$	$\dfrac{ML^2}{T^3}$
Time	$T,\, t$		second (s)	$\dfrac{1}{60}\text{min} = \dfrac{1}{3600}\text{h}$ $= \dfrac{1}{86400}\text{day}$	T
Velocity (velocity of light in vacuum = 300 ms⁻¹)	v	$\dfrac{\text{length}}{\text{time}}$	$\dfrac{\text{metre}}{\text{second}}$ (ms⁻¹)		$\dfrac{L}{T}$
Volume	V	length³	metre³ (m³)		L^3

Electrical units

Admittance	Y	$\dfrac{1}{\text{impedance}}$	mho ($\mho$)	$\dfrac{A}{V} = \dfrac{C^2}{J \cdot s} = S\dagger$	$\dfrac{I^2 T^3}{ML^2}$
Capacitance	C	$\dfrac{\text{charge}}{\text{potential}}$	farad (F)	$\dfrac{Q}{V} = \dfrac{C^2}{J} = \dfrac{A \cdot s}{V}$ $= 9 \times 10^{11}$ cm esu (cgs)	$\dfrac{I^2 T^4}{ML^2}$
Charge	Q, q	current $\times$ time	coulomb (C)	6.25×10^{18} electron charges $= A \cdot s$ $= 3 \times 10^9$ esu (cgs) $= 0.1$ emu (cgs)	IT
Charge density	ρ	$\dfrac{\text{charge}}{\text{volume}} = \nabla \cdot \boldsymbol{D}$	$\dfrac{\text{coulomb}}{\text{metre}^3}$ (Cm^{-3})	$\dfrac{A \cdot s}{m^3}$	$\dfrac{IT}{L^3}$
Conductance	G	$\dfrac{1}{\text{resistance}}$	mho ($\mho$)	$\dfrac{A}{V} = \dfrac{C^2}{J \cdot s} = S\dagger$	$\dfrac{I^2 T^3}{ML^2}$

(Contd...)

Table A.1 Fundamental, mechanical, electrical, and magnetic units (*Contd.*)

Name of dimension or quantity	Symbol	Description	SI unit and abbreviation	Equivalent unit	Dimension
Conductivity	σ	$\dfrac{1}{\text{resistivity}}$	$\dfrac{\text{mho}}{\text{metre}}\left(\mho\,\text{m}^{-1}\right)$	$\dfrac{1}{\Omega\text{m}}$	$\dfrac{I^2T^3}{ML^3}$
Current	I, i	$\dfrac{\text{charge}}{\text{time}}$	ampere (A)	$\dfrac{\text{C}}{\text{s}} = 3\times10^9\,\text{cgs (esu)}$ $= 0.1\,\text{cgs (emu)}$	I
Current density	J	$\dfrac{\text{current}}{\text{area}}$	$\dfrac{\text{ampere}}{\text{metre}^2}\left(\text{A}\cdot\text{m}^{-2}\right)$	$\dfrac{\text{C}}{\text{s}\cdot\text{m}^2}$	$\dfrac{I}{L^2}$
Dipole moment	$p(= ql)$	charge × length	coulomb·metre (C·m)	A·s·m	LIT
Emf	V	$\int \boldsymbol{E}\cdot d\boldsymbol{l}$	volt (V)	$\dfrac{\text{Wb}}{\text{s}} = \dfrac{\text{J}}{\text{C}}$	$\dfrac{ML^2}{IT^3}$
Energy density (electric)	w_e	$\dfrac{\text{energy}}{\text{volume}}$	$\dfrac{\text{joule}}{\text{metre}}\left(\text{Jm}^{-3}\right)$	$\dfrac{1}{100}\dfrac{\text{erg}}{\text{mm}^3}$	$\dfrac{M}{LT^2}$

Quantity	Symbol	Definition	Unit		Dimensions
Field intensity	E	$\dfrac{\text{potential}}{\text{length}} = \dfrac{\text{force}}{\text{charge}}$	$\dfrac{\text{volt}}{\text{metre}}$ (V m^{-1})	$\dfrac{\text{N}}{\text{C}} = \dfrac{\text{J}}{\text{C} \cdot \text{m}}$ $= \dfrac{1}{3} \times 10^{-4}\ \text{esu (cgs)}$ $= 10^{6}\ \text{emu (cgs)}$	$\dfrac{ML}{IT^3}$
Flux	ψ	$\text{charge} = \iint \boldsymbol{D} \cdot ds$	coulomb (C)	$\text{A} \cdot \text{s}$	IT
Flux density (displacement)	$\boldsymbol{D}$	$\dfrac{\text{charge}}{\text{area}}$	$\dfrac{\text{coulomb}}{\text{metre}^2}$ $(\text{C} \cdot \text{m}^{-2})$	$\dfrac{\text{A} \cdot \text{s}}{\text{m}^2} = \dfrac{\text{A}}{\text{m}^2 \text{s}^{-1}}$	$\dfrac{IT}{L^2}$
Impedance	Z	$\dfrac{\text{potential}}{\text{current}}$	ohm (Ω)	$\dfrac{\text{V}}{\text{A}}$	$\dfrac{ML^2}{I^2T^3}$
Linear charge density	ρ_L	$\dfrac{\text{charge}}{\text{length}}$	$\dfrac{\text{coulomb}}{\text{metre}}$ $(\text{C} \cdot \text{m}^{-1})$	$\dfrac{\text{A} \cdot \text{s}}{\text{m}}$	$\dfrac{IT}{L}$

(Contd...)

Table A.1 Fundamental, mechanical, electrical, and magnetic units (*Contd.*)

Name of dimension or quantity	Symbol	Description	SI unit and abbreviation	Equivalent unit	Dimension
Permittivity (dielectric constant) (for vacuum, $\varepsilon_0 = 8.85$ pF·m^{-1} $\approx 10^{-9}/36\pi$ F·m^{-1})	ε	$\dfrac{\text{capacitance}}{\text{length}}$	$\dfrac{\text{farad}}{\text{metre}}$ (F·m^{-1})	$\dfrac{\text{C}}{\text{Vm}}$	$\dfrac{I^2T^4}{ML^3}$
Polarization	$\boldsymbol{P}$	$\dfrac{\text{dipole moment}}{\text{volume}}$	$\dfrac{\text{coloumb}}{\text{metre}^2}$ (C·m^{-2})	$\dfrac{\text{A}\cdot\text{s}}{\text{m}^2}$	$\dfrac{IT}{L^2}$
Potential	V	$\dfrac{\text{work}}{\text{charge}}$	volt (V)	$\dfrac{\text{J}}{\text{C}} = \dfrac{\text{N}\cdot\text{m}}{\text{C}} = \dfrac{\text{W}\cdot\text{s}}{\text{C}}$ $= \dfrac{\text{W}}{\text{A}} = \dfrac{\text{Wb}}{\text{s}}$ $= \dfrac{1}{300}$ cgs esu $= 10^8$ cgs emu	$\dfrac{ML^2}{IT^3}$
Poynting vector	$\boldsymbol{S}$	$\dfrac{\text{power}}{\text{area}}$	$\dfrac{\text{watt}}{\text{metre}^2}$ (W·m^{-2})	$\dfrac{\text{J}}{\text{s}\cdot\text{m}^2}$	$\dfrac{M}{T^3}$

Quantity	Symbol	Definition	SI unit		Dimensions
Radiation intensity	P	$\dfrac{\text{power}}{\text{unit solid angle}}$	$\dfrac{\text{watt}}{\text{steradian}}$ $(W \cdot sr^{-1})$		$\dfrac{ML^2}{T^3}$
Reactance	X	$\dfrac{\text{potential}}{\text{current}}$	ohm (Ω)	$\dfrac{V}{A}$	$\dfrac{ML^2}{I^2T^3}$
Relative permittivity	ε_r	ratio $\dfrac{\varepsilon}{\varepsilon_0}$		dimensionless	
Resistance	R	$\dfrac{\text{potential}}{\text{current}}$	ohm (Ω)	$\dfrac{V}{A} = \dfrac{J \cdot s}{C^2}$ $= \dfrac{1}{9} \times 10^{-11}$ esu (cgs) $= 10^{-9}$ emu (cgs)	$\dfrac{ML^2}{I^2T^3}$
Resistivity	S	resistance $\times$ length $= \dfrac{1}{\text{conductivity}}$	ohm·metre $(\Omega \cdot m)$	$\dfrac{Vm}{A}$	$\dfrac{ML^3}{I^2T^3}$

(Contd...)

Table A.1 Fundamental, mechanical, electrical, and magnetic units (*Contd.*)

Name of dimension or quantity	Symbol	Description	SI unit and abbreviation	Equivalent unit	Dimension
Sheet-current density	K	$\dfrac{\text{current}}{\text{length}}$	$\dfrac{\text{ampere}}{\text{metre}}\,(\text{A}\cdot\text{m}^{-1})$	$\dfrac{\text{A}}{\text{m}^2}\times\text{m}$	$\dfrac{I}{L}$
Susceptance	B	$\dfrac{1}{\text{reactance}}$	mho ($\mho$)	$\dfrac{\text{A}}{\text{V}}=\text{S}$	$\dfrac{I^2 T^3}{ML^2}$
Wavelength	λ	length	metre (m)		L
Magnetic units					
Dipole moment (magnetic)	m $(=Q_m, l)$	pole strength $\times$ length $=$ current $\times$ area $=\dfrac{\text{torque}}{\text{magnetic flux density}}$	ampere $\cdot$ metre^2 ($\text{A}\cdot\text{m}^2$)	$\dfrac{\text{C}\cdot\text{m}^2}{\text{s}}$	IL^2
Energy density (magnetic)	w_m	$\dfrac{\text{energy}}{\text{volume}}$	$\dfrac{\text{joule}}{\text{metre}^3}\,(\text{J}\cdot\text{m}^{-3})$	$\dfrac{1}{100}\dfrac{\text{erg}}{\text{mm}^3}$	$\dfrac{M}{LT^2}$

Flux (magnetic)	ψ_m	$\iint \boldsymbol{B} \cdot d\boldsymbol{s}$	weber (Wb)	$V \cdot s = \dfrac{N \cdot m}{A}$ $= 10^8\,\text{Mx}$ cgs (emu)	$\dfrac{ML^2}{IT^2}$
Flux density	B	$\dfrac{\text{force}}{\text{pole}} = \dfrac{\text{force}}{\text{current moment}}$ $= \dfrac{\text{magnetic flux}}{\text{area}}$	tesla (T) $\dfrac{\text{weber}}{\text{metre}^2}$ $(\text{Wb} \cdot \text{m}^{-2})$	$\dfrac{V \cdot s}{m^2} = \dfrac{N}{A \cdot m}$ $= 10^4\,\text{G} \ddagger$ cgs (emu)	$\dfrac{M}{IT^2}$
Flux linkage	Λ	flux $\times$ turns	weber-turn (Wb-turn)		$\dfrac{ML^2}{IT^2}$
H field	H	$\dfrac{\text{mmf}}{\text{length}}$	$\dfrac{\text{ampere}}{\text{metre}}$ $(\text{A} \cdot \text{m}^{-1})$	$\dfrac{N}{Wb} = \dfrac{W}{V \cdot m}$ $= 4\pi \times 10^{-3}$ Oe [cgs] (emu) $= 400\pi$ gammas	$\dfrac{I}{L}$

(Contd...)

Table A.1 Fundamental, mechanical, electrical, and magnetic units (*Contd.*)

Name of dimension or quantity	Symbol	Description	SI unit and abbreviation	Equivalent unit	Dimension
Inductance	L	$\dfrac{\text{magnetic flux linkage}}{\text{current}}$	henry (H)	$\dfrac{\text{Wb}}{\text{A}} = \dfrac{\text{J}}{\text{A}^2} = \Omega \cdot \text{s}$ $= \dfrac{1}{9} \times 10^{-11} \text{ esu (cgs)}$ $= 10^9 \text{ cm emu (cgs)}$	$\dfrac{ML^2}{I^2 T^2}$
Magnetization (magnetic polarization)	M	$\dfrac{\text{magnetic moment}}{\text{volume}}$	$\dfrac{\text{ampere}}{\text{metre}}$ $(\text{A} \cdot \text{m}^{-1})$	$\dfrac{\text{A} \cdot \text{m}^2}{\text{m}^3} = \dfrac{\text{A} \cdot \text{m}}{\text{m}^2}$	$\dfrac{I}{L}$
Mmf	F	$\int \boldsymbol{H} \cdot d\boldsymbol{l}$	ampere-turn (A-turn)	$\dfrac{\text{C}}{\text{s}}$	I
Permeability (for vacuum, $\mu_0 = 400\pi \text{ nH m}^{-1}$)	μ	$\dfrac{\text{inductance}}{\text{length}}$	$\dfrac{\text{henry}}{\text{metre}}$ (Hm^{-1})	$\dfrac{\text{Wb}}{\text{A} \cdot \text{m}} = \dfrac{\text{V} \cdot \text{s}}{\text{A} \cdot \text{m}}$	$\dfrac{ML}{I^2 T^2}$

Permeance	P	$\dfrac{\text{magnetic flux}}{\text{mmf}}$ $= \dfrac{1}{\text{relutance}}$	$\dfrac{\text{Wb}}{\text{A}}$ henry (H)	$\dfrac{ML^2}{I^2T^2}$
Pole density	ρ_m	$\dfrac{\text{pole strength}}{\text{volume}}$ $= \dfrac{\text{current}}{\text{area}}$ $= \nabla \cdot \mathbf{H} = -\nabla \cdot \mathbf{M}$	$\dfrac{\text{ampere}}{\text{metre}^2}$ $(\text{A} \cdot \text{m}^{-2})$	$\dfrac{I}{L^2}$
Pole strength	Q_m, ρ_m	current $\times$ length $= \iiint \rho_m dv$	ampere-metre $(\text{A} \cdot \text{m})$ $\dfrac{\text{C} \cdot \text{m}}{\text{s}}$	IL
Potential (magnetic H)	U	$\int \mathbf{H} \cdot d\mathbf{l}$	ampere (A) $\dfrac{\text{J}}{\text{Wb}} = \dfrac{\text{W}}{\text{V}} = \dfrac{\text{C}}{\text{s}}$ $= \dfrac{4\pi}{10}\text{Gb\S}$ emu (cgs)	I

(Contd...)

Table A.1 Fundamental, mechanical, electrical, and magnetic units (*Contd.*)

Name of dimension or quantity	Symbol	Description	SI unit and abbreviation	Equivalent unit	Dimension
Relative permeability	μ_r	ratio $\dfrac{\mu}{\mu_0}$			Dimensionless
Reluctance	$\mathscr{R}$	$\dfrac{\text{mmf}}{\text{magnetic flux}}$ $= \dfrac{1}{\text{permeance}}$	$\dfrac{1}{\text{henry}}$ (H^{-1})	$\dfrac{\text{A}}{\text{Wb}}$	$\dfrac{I^2 T^2}{ML^2}$
Vector potential	A	current × permeability	$\dfrac{\text{Weber}}{\text{henry}}$ $(\text{Wb}\cdot\text{m}^{-1})$	$\dfrac{\text{HA}}{\text{m}} = \dfrac{\text{N}}{\text{A}}$	$\dfrac{ML}{IT^2}$

†S is the SI abbreviation for siemens, used often for mho.

‡Mx, G, and Oe are SI abbreviations for maxwell, gauss, and oersted.

§Gb is the SI abbreviation for gilbert.

APPENDIX B

Table B.1 Trigonometric, hyperbolic, logarithmic, and other relations

Trigonometric relations

$\sin (x \pm y) = \sin x \cos y \pm \cos x \sin y$

$\cos (x \pm y) = \cos x \cos y \pm \sin x \sin y$

$\sin (x + y) + \sin (x - y) = 2 \sin x \cos y$

$\cos (x + y) + \cos (x - y) = 2 \cos x \cos y$

$\sin (x + y) - \sin (x - y) = 2 \cos x \sin y$

$\cos (x + y) - \cos (x - y) = -2 \sin x \sin y$

$\sin 2x = 2 \sin x \cos x$

$\cos 2x = \cos^2 x - \sin^2 x = 2 \cos^2 x - 1 = 1 - 2 \sin^2 x$

$\cos x = 2\cos^2 \dfrac{1}{2}x - 1 = 1 - 2\sin^2 \dfrac{1}{2}x$

$\sin x = 2\sin\dfrac{1}{2}x\cos\dfrac{1}{2}x$

$\sin^2 x + \cos^2 x = 1$

$\tan (x + y) = \dfrac{\tan x + \tan y}{1 - \tan x \tan y}$

$\tan 2x = \dfrac{2\tan x}{1 - \tan^2 x}$

$\sin x = x - \dfrac{x^3}{3!} + \dfrac{x^5}{5!} - \dfrac{x^7}{7!} + ...$

$\cos x = 1 - \dfrac{x^2}{2!} + \dfrac{x^4}{4!} - \dfrac{x^6}{6!} + ...$

$\tan x = x + \dfrac{x^3}{3} + \dfrac{2x^5}{15} + \dfrac{17x^7}{315} + \dfrac{62x^9}{2835} + ...$

Hyperbolic relations

$\sinh x = \dfrac{e^x - e^{-x}}{2} = x + \dfrac{x^3}{3!} + \dfrac{x^5}{5!} + \dfrac{x^7}{7!} + ...$

$\cosh x = \dfrac{e^x + e^{-x}}{2} = 1 + \dfrac{x^2}{2!} + \dfrac{x^4}{4!} + \dfrac{x^6}{6!} + ...$

$\tanh x = \dfrac{\sinh x}{\cosh x}$

Table B.1 Trigonometric, hyperbolic, logarithmic, and other relations (*Contd.*)

Hyperbolic relations

$$\coth x = \frac{\cosh x}{\sinh x} = \frac{1}{\tanh x}$$

$\sinh (x \pm jy) = \sinh x \cos y \pm j \cosh x \sin y$

$\cosh (x \pm jy) = \cosh x \cos y \pm j \sinh x \sin y$

$$\left. \begin{array}{l} \cosh(jx) = \dfrac{1}{2}\left(e^{+jx} + e^{-jx}\right) = \cos x \\[2mm] \sinh(jx) = \dfrac{1}{2}\left(e^{+jx} - e^{-jx}\right) = j \sin x \end{array} \right\} \text{ de Moivre's theorem}$$

$e^{\pm jx} = \cos x \pm j \sin x$

$$e^{\pm jx} = 1 \pm jx - \frac{x^2}{2!} \mp j\frac{x^3}{3!} + \frac{x^4}{4!} \pm j\frac{x^5}{5} - \ldots$$

$e^{x} = \cosh x + \sinh x$

$e^{-x} = \cosh x - \sinh x$

$$e^{x} = 1 + x + \frac{x^2}{2!} + \frac{x^3}{3!} + \frac{x^4}{4!} + \ldots$$

$\cosh x = \cos jx$

$j \sinh x = \sin jx$

$$\tanh(x \pm jy) = \frac{\sinh 2x}{\cosh 2x + \cos 2y} \pm j \frac{\sin 2y}{\cosh 2x + \cos 2y}$$

$$\coth(x \pm jy) = \frac{\sinh 2x}{\cosh 2x - \cos 2y} \pm j \frac{\sin 2y}{\cosh 2x - \cos 2y}$$

Logarithmic relations

$\log_{10} x = \log x$ common logarithm

$\log_{e} x = \ln x$ natural logarithm

$\log_{10} x = 0.4343 \log_{e} x = 0.4343 \ln x$

   $\ln x = \log_{e} x = 2.3026 \log_{10} x$

     $e = 2.71828$

    dB $= 10 \log (\text{power ratio}) = 20 \log (\text{voltage ratio})$

1 Np$(\text{voltage attenuation}) = \dfrac{1}{e} = 0.368(\text{voltage}) = -8.68$ dB

Approximation formulas for small quantities

(δ is a small quantity compared with unit)

$(1 \pm \delta)^2 = 1 \pm 2\delta$

$(1 \pm \delta)^n = 1 \pm n\delta$

$\sqrt{1+\delta} = 1 + \frac{1}{2}\delta$

$\dfrac{1}{\sqrt{1+\delta}} = 1 - \frac{1}{2}\delta$

$e^\delta = 1 + \delta$

$\ln(1 + \delta) = \delta$

$J_n(\delta) = \dfrac{\delta^n}{n!2^n}$ for $|\delta| \ll 1$

where J_n is Bessel function of order n. Thus

$J_1(\delta) = \dfrac{\delta}{2}$

Series

Binomial:

$$(x+y)^n = x^n + nx^{n-1}y + \frac{n(n-1)}{2!}x^{n-2}y^2 + \frac{n(n-1)(n-2)}{3!}x^{(n-3)}y^3 + \ldots$$

Taylor's:

$$f(x+y) = f(x) + \frac{df(x)}{dx}\frac{y}{1} + \frac{d^2f(x)}{dx^2}\frac{y^2}{2!} + \frac{d^3f(x)}{dx^3}\frac{y^3}{3!} + \ldots$$

Solution of quadratic equation

If $ax^2 + bx + c = 0$, then

$$x = \frac{-b \pm \sqrt{b^2 - 4ac}}{2a}$$

Gradient, divergence, and curl in rectangular, cylindrical, and spherical coordinates

Rectangular coordinates

$$\nabla f = \hat{x}\frac{\partial f}{\partial x} + \hat{y}\frac{\partial f}{\partial x} + \hat{z}\frac{\partial f}{\partial x}$$

Table B.1 Trigonometric, hyperbolic, logarithmic, and other relations (*Contd.*)

Rectangular coordinates

$$\nabla \cdot A = \frac{\partial A_x}{\partial x} + \frac{\partial A_y}{\partial y} + \frac{\partial A_z}{\partial z}$$

$$\nabla \times A = \hat{x}\left(\frac{\partial A_z}{\partial y} - \frac{\partial A_y}{\partial z}\right) + \hat{y}\left(\frac{\partial A_x}{\partial z} - \frac{\partial A_z}{\partial x}\right) + \hat{z}\left(\frac{\partial A_y}{\partial x} - \frac{\partial A_x}{\partial y}\right) \begin{vmatrix} \hat{x} & \hat{y} & \hat{z} \\ \frac{\partial}{\partial x} & \frac{\partial}{\partial y} & \frac{\partial}{\partial z} \\ A_x & A_y & A_z \end{vmatrix}$$

Cylindrical coordinates

$$\nabla f = \hat{r}\frac{\partial f}{\partial r} + \hat{\Phi}\frac{1}{r}\frac{\partial f}{\partial \phi} + \hat{z}\frac{\partial f}{\partial z}$$

$$\nabla \cdot A = \frac{1}{r}\frac{\partial}{\partial r}rA_r + \frac{1}{r}\frac{\partial A_\phi}{\partial \phi} + \frac{\partial A_z}{\partial z}$$

$$\nabla \times A = \hat{r}\left(\frac{1}{r}\frac{\partial A_z}{\partial \phi} - \frac{\partial A_\phi}{\partial z}\right) + \hat{\Phi}\left(\frac{\partial A_r}{\partial z} - \frac{\partial A_z}{\partial r}\right) + \hat{z}\frac{1}{r}\left(\frac{\partial}{\partial r}rA_\phi - \frac{\partial A_r}{\partial \phi}\right)$$

$$= \begin{vmatrix} \hat{r}\frac{1}{r} & \hat{\Phi} & \hat{z}\frac{1}{r} \\ \frac{\partial}{\partial r} & \frac{\partial}{\partial \phi} & \frac{\partial}{\partial z} \\ A_r & rA_\phi & A_z \end{vmatrix}$$

Spherical coordinates

$$\nabla f = \hat{r}\frac{\partial f}{\partial r} + \hat{\theta}\frac{1}{r}\frac{\partial f}{\partial \theta} + \hat{\Phi}\frac{1}{r\sin\theta}\frac{\partial f}{\partial \phi}$$

$$\nabla \cdot A = \frac{1}{r^2}\frac{\partial}{\partial r}r^2 A_r + \frac{1}{r\sin\theta}\frac{\partial}{\partial \theta}(A_\theta \sin\theta) + \frac{1}{r\sin\theta}\frac{\partial A_\phi}{\partial \phi}$$

$$\nabla \times A = \hat{r}\frac{1}{r\sin\theta}\left(\frac{\partial}{\partial \theta}(A_\phi \sin\theta) - \frac{\partial A_\theta}{\partial \phi}\right) + \hat{\theta}\frac{1}{r}\left(\frac{1}{\sin\theta}\frac{\partial A_r}{\partial \phi} - \frac{\partial}{\partial r}rA_\phi\right)$$

$$+ \hat{\Phi}\frac{1}{r}\left(\frac{\partial}{\partial r}rA_\theta - \frac{\partial A_r}{\partial \theta}\right)$$

APPENDIX C

Table C.1 Glyphs (nonalphabetic pictograph symbols)

Symbol	Definition
$=$	Equal to
$\sim$ or $\approx$	Approximately equal to
$\cong$	Nearly equal to
$\equiv$	Identical with or by definition
$\neq$	Not equal to
$\propto$	Proportional to
$\%$	Percent
$\rightarrow$	Approaches
$<$	Less than
$>$	Greater than
$\leq$	Less than or equal to
$\geq$	Greater than or equal to
$<<$	Much less than
$>>$	Much greater than
∞	Infinity
$\therefore$	Therefore
$!$	Factorial
$\sqrt{}$	Square root
$\|\ \|$	Absolute value
Σ	Summation sign
$\int$	Integral sign
$\oint$	Line integral around a closed path
$\iint$ or $\int_s$	Surface integral
$\oiint$ or $\oint_s$	Surface integral completely enclosing a volume
$\iiint$ or $\int_v$	Volume integral